On Dynamical Poisson Groupoids I

of the
American Mathematical Society

Number 824

On Dynamical Poisson Groupoids I

Luen-Chau Li
Serge Parmentier

March 2005 • Volume 174 • Number 824 (end of volume) • ISSN 0065-9266

American Mathematical Society
Providence, Rhode Island

2000 *Mathematics Subject Classification.* Primary 53D17; Secondary 58H05.

Library of Congress Cataloging-in-Publication Data

Li, Luen-Chau, 1954–
   On dynamical Poisson groupoids I / Luen-Chau Li, Serge Parmentier.
     p. cm. — (Memoirs of the American Mathematical Society, ISSN 0065-9266 ; no. 824)
   "Volume 174, number 824 (end of 4 numbers)."
   Includes bibliographical references.
   1SBN 0-8218-3673-0 (alk. paper)
     1. Poisson manifolds. 2. Pseudogroups. I. Parmentier, Serge, 1961– II. Title. III. Series.

QA3.A57 no. 824 2005
510 s—dc22
[512′.2]
                                                                     2004062689

Memoirs of the American Mathematical Society

This journal is devoted entirely to research in pure and applied mathematics.

Subscription information. The 2005 subscription begins with volume 173 and consists of six mailings, each containing one or more numbers. Subscription prices for 2005 are $606 list, $485 institutional member. A late charge of 10% of the subscription price will be imposed on orders received from nonmembers after January 1 of the subscription year. Subscribers outside the United States and India must pay a postage surcharge of $31; subscribers in India must pay a postage surcharge of $43. Expedited delivery to destinations in North America $35; elsewhere $130. Each number may be ordered separately; *please specify number* when ordering an individual number. For prices and titles of recently released numbers, see the New Publications sections of the *Notices of the American Mathematical Society*.

Back number information. For back issues see the *AMS Catalog of Publications*.

Subscriptions and orders should be addressed to the American Mathematical Society, P. O. Box 845904, Boston, MA 02284-5904, USA. *All orders must be accompanied by payment.* Other correspondence should be addressed to 201 Charles Street, Providence, RI 02904-2294, USA.

Copying and reprinting. Individual readers of this publication, and nonprofit libraries acting for them, are permitted to make fair use of the material, such as to copy a chapter for use in teaching or research. Permission is granted to quote brief passages from this publication in reviews, provided the customary acknowledgment of the source is given.

Republication, systematic copying, or multiple reproduction of any material in this publication is permitted only under license from the American Mathematical Society. Requests for such permission should be addressed to the Acquisitions Department, American Mathematical Society, 201 Charles Street, Providence, Rhode Island 02904-2294, USA. Requests can also be made by e-mail to reprint-permission@ams.org.

Memoirs of the American Mathematical Society is published bimonthly (each volume consisting usually of more than one number) by the American Mathematical Society at 201 Charles Street, Providence, RI 02904-2294, USA. Periodicals postage paid at Providence, RI. Postmaster: Send address changes to Memoirs, American Mathematical Society, 201 Charles Street, Providence, RI 02904-2294, USA.

© 2005 by the American Mathematical Society. All rights reserved.
This publication is indexed in *Science Citation Index*®, *SciSearch*®, *Research Alert*®, *CompuMath Citation Index*®, *Current Contents*®/*Physical, Chemical & Earth Sciences*.
Printed in the United States of America.

∞ The paper used in this book is acid-free and falls within the guidelines established to ensure permanence and durability.
Visit the AMS home page at http://www.ams.org/

10 9 8 7 6 5 4 3 2 1   10 09 08 07 06 05

Contents

Chapter 1. Introduction — 1

Chapter 2. A class of biequivariant Poisson groupoids — 5
   2.1. Preliminaries — 5
   2.2. Trivial Lie groupoids in $\mathcal{C}_U$ — 8

Chapter 3. Duality — 17
   3.1. Duality of Poisson groupoids — 17
   3.2. The dual of a dynamical Poisson groupoid — 19

Chapter 4. An explicit case study of duality — 26

Chapter 5. Coboundary dynamical Poisson groupoids - the constant r-matrix case — 34
   5.1. The dual Poisson groupoid — 34
   5.2. Construction of the associated symplectic double groupoid — 42

Appendix — 65
   A.1. Proof of Proposition 2.2.3 — 65
   A.2. Proof of Theorem 2.2.5 (b) — 66
   A.3. Proof of Proposition 3.2.1 — 67
   A.4. Proof of the coisotropy in Theorem 5.1.4 — 69

Bibliography — 71

Abstract

We address the question of duality for the dynamical Poisson groupoids of Etingof and Varchenko over a contractible base. We also give an explicit description for the coboundary case associated with the solutions of the classical dynamical Yang-Baxter equation on simple Lie algebras as classified by the same authors. Our approach is based on the study of a class of Poisson structures on trivial Lie groupoids within the category of biequivariant Poisson manifolds. In the former case, it is shown that the dual Poisson groupoid of such a dynamical Poisson groupoid is isomorphic to a Poisson groupoid (with trivial Lie groupoid structure) within this category. In the latter case, we find that the dual Poisson groupoid is also of dynamical type modulo Poisson groupoid isomorphisms. For the coboundary dynamical Poisson groupoids associated with constant r-matrices, we give an explicit construction of the corresponding symplectic double groupoids. In this case, the symplectic leaves of the dynamical Poisson groupoid are shown to be the orbits of a Poisson Lie group action.

Received by the editor September 18, 2002.
2000 *Mathematics Subject Classification*. Primary 53D17; Secondary 58H05.
Key words and phrases. Poisson groupoids, duality, classical dynamical r-matrices, symplectic double groupoids.

CHAPTER 1

Introduction

The classical dynamical Yang-Baxter equation (CDYBE) is a functional differential equation which first appeared in the context of Wess-Zumino-Witten conformal field theory [**BDF**], [**F**]. Subsequently, its geometric meaning was unraveled by Etingof and Varchenko in the fundamental paper [**EV**]. While the solutions of the classical Yang-Baxter equation are related to Poisson Lie groups [**D**], the authors in [**EV**] showed that an appropriate geometrical setting for the CDYBE is that of a special class of Poisson groupoids (as defined in [**W1**]), the so-called coboundary dynamical Poisson groupoids. Given a Lie group G, a Lie subgroup $H \subset G$, an Ad_H^* invariant open set $U \subset \mathfrak{h}^*$ (here $\mathfrak{h}^*$ is the dual of $\mathfrak{h} = Lie(H)$), and a solution of the CDYBE, Etingof and Varchenko constructed a Poisson bracket on the trivial Lie groupoid $X = U \times G \times U$ compatible with its groupoid structure. This Poisson bracket intertwines terms which are responsible for left and right inclusions of the restricted symplectic cotangent $H \times U$ into X together with a Sklyanin-like term on G. In addition, the authors in [**EV**] identified an appropriate abstract context in which to view these objects as the category of H-bi-equivariant Poisson manifolds $\mathcal{C}_U$.

It is classical that the study of Poisson Lie groups relies in an essential way on duality and the construction of doubles [**D**], [**STS**], [**LW1**]. For Poisson groupoids, the notion of duality was introduced by Weinstein in [**W1**], and was developed by Mackenzie and Xu in [**MX1**],[**MX2**]. In the same paper [**W1**], Weinstein also introduced the notion of symplectic double groupoids (see also [**M2**]), and described a program for showing that, at least locally, Poisson groupoids in duality arise as the base of a symplectic double groupoid.

In order to state our objectives and results, let us begin by recalling that a symplectic groupoid is a pair (Γ, Π), consisting of a Lie groupoid Γ together with a non-degenerate Poisson structure Π, in such a way that the graph of the multiplication map is a Lagrangian submanifold of $\Gamma \times \Gamma \times \overline{\Gamma}$ [**W2**],[**K**]. It is a classical fact that Poisson structures can be understood at least locally by the notion of symplectic groupoids. On the other hand, double groupoids are intrinsically complicated objects introduced by Ehresmann [**E**] in the 1960's and have found usage in category theory [**E**], homotopy theory [**BH**], differential geometry [**P**], and Poisson groups [**M3**], [**LW2**]. By definition, a double Lie groupoid is a quadruple $(\mathcal{S}; \mathcal{H}, \mathcal{V}, B)$ where $\mathcal{H}$ and $\mathcal{V}$ are Lie groupoids over B, and $\mathcal{S}$ is equipped with two Lie groupoid structures, a horizontal structure with base $\mathcal{V}$, and a vertical structure with base $\mathcal{H}$, such that the structure maps of each groupoid structure on $\mathcal{S}$ are morphisms with respect to the other. Finally, a symplectic double groupoid is a double Lie groupoid $(\mathcal{S}; \mathcal{H}, \mathcal{V}, B)$ in which $\mathcal{S}$ is equipped with a symplectic structure such that both $\mathcal{S} \rightrightarrows \mathcal{V}$ and $\mathcal{S} \rightrightarrows \mathcal{H}$ are symplectic groupoids. Note that for the case of Poisson Lie groups, the program in [**W1**] which we mentioned above has been carried out globally in [**LW2**]. Thus a Poisson Lie group and its dual are the bases of a symplectic groupoid.

This work is the first part of a series to understand the geometry of dynamical Poisson groupoids. Our goal here is three-fold. First of all, for a general dynamical Poisson groupoid $X = U \times G \times U$ (not necessarily of coboundary type), we would like to characterize certain properties of its (global) dual Poisson groupoid in a simple nontrivial case in which its existence is guaranteed. In this connection, we should point out that in contrast to (finite dimensional) Lie algebras, not all Lie algebroids can be integrated to Lie groupoids [**AM**]. For (finite dimensional) general Lie algebroids, the necessary and sufficient condition for integrability was only obtained quite recently in [**CF**]. Thus we work at the outset with the class $\mathcal{C}_*$ of Poisson groupoids $X = U \times G \times U$ (with the trivial Lie groupoid structure) which admits a (base preserving) Poisson groupoid morphism $I : H \times U \longrightarrow X$, where U is Ad_H^* invariant and contractible. If X is a dynamical Poisson groupoid in $\mathcal{C}_*$, the corresponding Lie algebroid dual $A(X)^*$ must be transitive, i.e., the anchor map is a surjective submersion. Consequently, we can invoke a general theorem of Mackenzie [**M1**], according to which $A(X)^* \simeq TU \oplus (U \times \mathfrak{g}')$, where the latter is the trivial Lie algebroid over U and $\mathfrak{g}'$ is a typical fiber of the adjoint bundle of $A(X)^*$. As a result, $A(X)^*$ integrates to a unique global Lie groupoid X^* isomorphic to the trivial Lie groupoid $U \times G' \times U$, where G' is the connected and simply connected Lie group with $Lie(G') = \mathfrak{g}'$. Thus the existence of the dual Poisson groupoid is not an issue. Our main result in this direction (Theorem 3.2.4) is the following: if X is a dynamical Poisson groupoid in $\mathcal{C}_*$, then its dual Poisson groupoid X^* is isomorphic to a Poisson groupoid $(U \times G' \times U, \{\,,\,\}_{U \times G' \times U})$ in $\mathcal{C}_*$. In particular, the Poisson structure $\{\,,\,\}_{U \times G' \times U}$ is uniquely determined by a (unique) Poisson groupoid morphism $I' : H \times U \longrightarrow U \times G' \times U$ and a unique groupoid 1-cocycle P' on $U \times G' \times U$. The proof of this theorem consists of two steps: in the first step, we establish the existence of the Poisson groupoid morphism I'; while the second step involves a careful analysis of the form of the Poisson bracket for a Poisson groupoid in $\mathcal{C}_*$(Theorem 2.2.5). As a corollary of Theorem 3.2.4, we obtain via Poisson reduction a reduced duality diagram for the Poisson quotients $G/H \times U$ and $G'/H \times U$ and for the vertex Lie algebras $\mathfrak{g}$ and $\mathfrak{g}'$. In the special case when $\mathfrak{h}^* = 0$, this duality diagram is just the well-known diagram of Drinfel'd for Poisson Lie groups.

In [**EV**], extending Belavin and Drinfel'd's classic paper [**BD**], Etingof and Varchenko obtained a classification of solutions of the CDYBE for pairs $(\mathfrak{g}, \mathfrak{h})$ of Lie algebras, where $\mathfrak{g}$ is simple and $\mathfrak{h} \subset \mathfrak{g}$ is a Cartan subalgebra. These solutions of the CDYBE are parametrized by subsets S of a simple system of roots Δ^s and closed meromorphic two-forms on $\mathfrak{h}^*$.

Our second objective is to give an explicit study of duality for the coboundary dynamical Poisson groupoids associated with this class of dynamical r-matrices. Note that in this case, the base U (where the r-matrix is analytic) is neither contractible nor simply-connected. We proceed in two steps. To start with, we construct (see Theorem 4.4) an explicit trivialization of the Lie algebroid dual $A(X)^*$ of the (full) coboundary Poisson groupoid $X = U \times G \times U$. This, in particular, establishes the integrability of $A(X)^*$ as a Lie algebroid. Then an argument similar to that of Theorem 3.2.4 applied to any connected and simply connected open subset U' of U shows (see Theorem 4.5) that the dual Poisson groupoid of $U' \times G \times U'$ is isomorphic to a dynamical Poisson groupoid $U' \times G' \times U'$. Here, the vertex Lie group G' is a semi-direct product $L_S \ltimes I_S$ where $L_S \subset G$ is the Levi factor and I_S is a normal Lie subgroup containing the product $N_S^+ \times N_S^-$ of unipotent radicals. More importantly, the Poisson bracket is uniquely determined by the value of a Lie groupoid 1-cocycle $P' : U' \times G' \times U' \longrightarrow L(\mathfrak{g}'^*, \mathfrak{g}')$ whose partial derivatives are explicitly given in terms of (the bilinear part of) the Lie-Yamaguti data of the reductive pair $(\mathfrak{g}, \mathfrak{h})$.

Our final objective in this paper is to understand how to construct symplectic double groupoids for the coboundary dynamical case in the special instance where the r-matrix is constant. For this class of coboundary dynamical Poisson groupoids, the base is $\mathfrak{h}^*$ and so Theorem 3.2.4 applies. However, from the point of view of constructing the symplectic double groupoids, it is more natural (and considerably simpler) to work directly with the dual Poisson groupoid whose Lie algebroid is $T^*\mathfrak{h}^* \times \mathfrak{g}^*$. Since we have a constant r-matrix, the Lie group G equipped with the Sklyanin bracket is a Poisson Lie group (for simplicity, we assume G is complete) and as it turns out, the dual Poisson groupoid of X is given by $X^* = H \times \mathfrak{h}^* \times G^*$ with appropriate structure maps (G^* is the dual Poisson group of G) and the Poisson structure is a product structure (Theorem 5.1.4). The construction of a symplectic double groupoid having X and X^* as side groupoids proceeds via a number of steps (Proposition 5.2.3, Corollary 5.2.6, Corollary 5.2.8, Theorem 5.2.10 and 5.2.13). First of all, we show X and X^* form a matched pair of Lie groupoids. The upshot of this is that X and X^* act on each other via groupoid actions and give rise to a vacant double Lie groupoid $(\mathcal{S}_{vac}; X^*, X, \mathfrak{h}^*)$. However, this is not the correct underlying double Lie groupoid of the sought-for symplectic double groupoid (in contrast to the Poisson group case). In the second step of the construction, we extend the Lie groupoids X and X^* to the product groupoids $X_e^* = X^* \times H \rightrightarrows H \times \mathfrak{h}^*$ and $X_e = (H \times H) \times X \rightrightarrows H \times \mathfrak{h}^*$ ($H \times H \longrightarrow H$ is the coarse groupoid). Then we show that there is a left action of X_e^* on X and a right action of X_e on X^*. The corresponding action groupoids $\mathcal{S} \simeq X_e^* \ltimes X \rightrightarrows X$ and $\mathcal{S} \simeq X^* \rtimes X_e \rightrightarrows X^*$ then give the horizontal structure and the vertical structure respectively of a nonvacant double Lie groupoid $(\mathcal{S}; X^*, X, \mathfrak{h}^*)$ which has $(\mathcal{S}_{vac}; X^*, X, \mathfrak{h}^*)$ as a double Lie subgroupoid. Finally, we show that the double Lie groupoid $(\mathcal{S}; X^*, X, \mathfrak{h}^*)$ where $\mathcal{S}$ is equipped with an appropriate symplectic structure is a desired symplectic double groupoid. We would like to point out that the actions of the extended Lie groupoids on the unextended ones obey a number of properties (Proposition 5.2.9) which are important in showing that $(\mathcal{S}; X^*, X, \mathfrak{h}^*)$ is a double Lie groupoid. The reader should contrast these properties with actions via 'twisted automorphisms' (Proposition 5.2.4, [M3], [LW1]). As an application and amplification of this result, we show the existence of a natural Poisson Lie group structure on the set $H \times H \times G^*$ such that the symplectic leaves of $(X, \{,\}_X)$ are the orbits of a Poisson action of $H \times H \times G^*$ on X (Theorem 5.2.29). Finally, we use this result to describe the symplectic leaves of a natural Poisson quotient associated with X.

The paper is organized as follows. In Section 2, we begin by giving some background material which we recall here for the convenience of the reader. The rest of Section 2 is devoted to the description of all Poisson groupoids $(X, \{,\})$ which admit a Poisson groupoid morphism $I : H \times U \longrightarrow X = U \times G \times U$, where

X is the trivial Lie groupoid over a connected base U. Section 3 is concerned with Poisson groupoids in duality with dynamical Poisson groupoids over a contractible base U. It also treats duality diagrams for the Poisson quotients mentioned earlier and for the vertex Lie algebras. In Section 4, we consider the coboundary dynamical Poisson groupoids associated with a class of solutions of (CDYBE) for pairs $(\mathfrak{g}, \mathfrak{h})$ of Lie algebras, where $\mathfrak{g}$ is simple, and $\mathfrak{h}$ is a Cartan subalgebra of $\mathfrak{g}$ [**EV**]. Here, we obtain a more refined description of the dual Poisson groupoid. Finally, Section 5 treats the coboundary dynamical Poisson groupoids in the constant r-matrix case in detail. We begin with an explicit description of the dual Poisson groupoid X^* whose Lie algebroid is $T^*\mathfrak{h}^* \times \mathfrak{g}^*$. Then we move on to the construction of a symplectic double groupoid having X and X^* as side groupoids. We conclude the paper by describing the symplectic leaves of $(X, \{\,,\,\})$ as well as a Poisson quotient associated with X.

We will address the construction of symplectic double groupoids for the general dynamical case, together with its relationship to other works (in particular [**LWX**]) in a sequel to this paper. For the relevance of coboundary dynamical Poisson groupoids and coboundary dynamical Lie algebroids in integrable systems, we refer the reader to [**HM**],[**LX1**],[**LX2**],[**L1**],[**L2**] and forthcoming publications.

Acknowledgements. L.-C. Li would like to thank the members of the Institut G. Desargues for hospitality and CNRS support (UMR 5028) during his visits to Université Lyon 1. He is also indebted to Eyal Markman for helpful comments on Remark 5.2.33. S. Parmentier thanks R. Pujol for his observation in Remark 2.2.7 (d).

CHAPTER 2

A Class of Biequivariant Poisson Groupoids

2.1. Preliminaries

In this preliminary subsection, we recall some of the basic concepts and constructs which we will use in this paper (other results will be recalled when needed).

Let Γ be a Lie groupoid over B (see [**CdSW**],[**M1**] for details), with target and source maps $\alpha, \beta : \Gamma \longrightarrow B$, and multplication map $m : \Gamma * \Gamma \longrightarrow B$ defined on the set of composable pairs $\Gamma * \Gamma := \{(x, y) \mid \beta(x) = \alpha(y)\}$. We will denote the unit section by $\epsilon : B \longrightarrow \Gamma$, and the inversion map by $i : \Gamma \longrightarrow \Gamma$.

Definition 2.1.1 [W1]. (Poisson groupoid.) *A Lie groupoid Γ equipped with a Poisson structure Π is called a Poisson groupoid if and only if the graph of the multiplication map*

$$Gr(m) \subset \Gamma \times \Gamma \times \overline{\Gamma}$$

is a coisotropic submanifold, i.e. if and only if

$$(\Pi \oplus \Pi \oplus -\Pi)(\omega, \omega') = 0, \quad \forall \omega, \omega' \in (T(Gr(m)))^\perp \subset T^*(\Gamma \times \Gamma \times \Gamma).$$

Γ *is called a symplectic groupoid if Π is non degenerate with $Gr(m)$ a Lagrangian submanifold.*

In both cases, the Poisson structure and the groupoid structure are said to be compatible.

Let G be a connected Lie group, $H \subset G$ a connected Lie subgroup with respective Lie algebras $\mathfrak{g}$ and $\mathfrak{h}$ and let $U \subset \mathfrak{h}^*$ be a connected Ad_H^*- invariant open subset. In [**EV**], Etingof and Varchenko introduced the category $\mathcal{C}_U$ of biequivariant Poisson manifolds over U as follows.

An object in $\mathcal{C}_U$ is a Poisson manifold $(X, \{ , \}_X)$ equipped with commuting left Hamiltonian H-action ϕ^- and right Hamiltonian H-action ϕ^+ with U-valued Ad_H^*-equivariant momentum maps $j_\pm : X \longrightarrow U$ satisfying the polarity condition

$$\{j_+^* \varphi, j_-^* \psi\}_X = 0, \quad \text{for all } \varphi, \psi \in C^\infty(U).$$

A morphism in $\mathcal{C}_U$ between $(X, \{ , \}_X)$ and $(X', \{ , \}_{X'})$ is an equivariant Poisson map $\sigma : X \longrightarrow X'$ such that $j'_\pm \circ \sigma = j_\pm$.

Definition 2.1.2 [EV]. (Poisson groupoid in $\mathcal{C}_U$.) *A Poisson manifold $X \in \mathcal{C}_U$ is a Poisson groupoid in $\mathcal{C}_U$ iff it is equipped with a compatible groupoid structure over U such that $\alpha = j_-$, $\beta = j_+$.*

Example 2.1.3. (The Hamiltonian unit.) The most basic (but not simplest) symplectic groupoid in $\mathcal{C}_U$ is the (restricted) Hamiltonian unit $H \times U$ equipped with the non-degenerate bracket

$$\{f,g\}(h,p) = - <D'g, \delta f> + <D'f, \delta g> - <p, [\delta f, \delta g]>,$$

$(<D'f, Z> = \frac{d}{dt}_{|t=0} f(he^{tZ}, p), \quad <\delta f, \lambda> = \frac{d}{dt}_{|t=0} f(h, p+t\lambda), Z \in \mathfrak{h}, \lambda \in \mathfrak{h}^*)$
the H-actions

$$\phi_k^-(h,p) = (kh, p), \quad \phi_k^+(h,p) = (hk, Ad_k^* p),$$

and the action groupoid structure

$$\alpha_0(h,p) = j_-(h,p) = Ad_{h^{-1}}^* p, \quad \beta_0(h,p) = j_+(h,p) = p$$
$$(h, j_-(k,q)) \cdot (k,q) = (hk, q), \quad \epsilon(q) = (1, q), \quad i(h,p) = (h^{-1}, Ad_{h^{-1}}^* p).$$

If $U = \mathfrak{h}^*$, this is clearly isomorphic to the cotangent symplectic groupoid T^*H [**W2**] under the trivialization map.

We now recall a fundamental construction of [**EV**] which interprets dynamical r-matrices in terms of Poisson groupoids.

Let $\iota : \mathfrak{h} \longrightarrow \mathfrak{g}$ be the Lie inclusion. We say that a smooth map $R : U \longrightarrow L(\mathfrak{g}^*, \mathfrak{g})$ (here and henceforth we denote by $L(\mathfrak{g}^*, \mathfrak{g})$ the set of linear maps from $\mathfrak{g}^*$ to $\mathfrak{g}$) is a classical dynamical r-matrix if and only if it is pointwise skew symmetric :

$$<R(q)(A), B> = - <A, R(q)B>,$$

and satisfies the classical dynamical Yang- Baxter condition

$$\begin{aligned} dR(q)\iota^*A\,(B) - dR(q)\iota^*B\,(A) + \iota d <R(q)A, B> \\ - [R(q)A, R(q)B] - R(q)ad^*_{R(q)(A)}B + R(q)ad^*_{R(q)B}A = \chi(A,B), \end{aligned} \quad (2.1.1)$$

where $\chi : \mathfrak{g}^* \times \mathfrak{g}^* \longrightarrow \mathfrak{g}$ is $ad_\mathfrak{g}$-invariant, i.e.

$$ad_X\, \chi(A,B) + \chi(ad_X^* A, B) + \chi(A, ad_X^* B) = 0,$$

for all $A, B \in \mathfrak{g}^*, X \in \mathfrak{g}$, and all $q \in U$.

The dynamical r-matrix is said to be $ad_\mathfrak{h}^*$-equivariant if and only if

$$dR(q)ad_Z^* q + R(q)ad_{\iota(Z)}^* + ad_{\iota(Z)} R(q) = 0, \quad (2.1.2)$$

for all $Z \in \mathfrak{h}$, and all $q \in U$.

Note that if $\chi(A,B) = 0$ in (2.1.1), the resulting equation is called the classical dynamical Yang-Baxter equation (CDYBE) [**F**]. On the other hand, if $\chi(A,B) = [T(A), T(B)]$ for some nonzero symmetric map $T : \mathfrak{g}^* \longrightarrow \mathfrak{g}$ with $ad_X \circ T + T \circ ad_X^* = 0$, the resulting equation is called the modified dynamical Yang-Baxter equation (mDYBE).

Let $X = U \times G \times U$. For $f \in C^\infty(X)$, define its partial derivatives and its left and right gradients (with respect to G) by

$$<\delta_1 f, \lambda> = \frac{d}{dt}_{|t=0} f(p + t\lambda, x, q), \quad <\delta_2 f, \lambda> = \frac{d}{dt}_{|t=0} f(p, x, q + t\lambda), \lambda \in \mathfrak{h}^*$$

$$<Df, X> = \frac{d}{dt}_{|t=0} f(p, e^{tX}x, q), \quad <D'f, X> = \frac{d}{dt}_{|t=0} f(p, xe^{tX}, q), X \in \mathfrak{g}.$$

We will equip X with the trivial Lie groupoid structure over U with structure maps

$$\alpha(p,x,q) = p, \ \beta(p,x,q) = q, \ \epsilon(q) = (q,1,q), \ i(p,x,q) = (q,x^{-1},p)$$
$$m((p,x,q),(q,y,r)) = (p,xy,r). \tag{2.1.3}$$

The following theorem gives the Poisson groupoid analog of coboundary Poisson Lie groups (in the context of trivial Lie groupoids over U).

Theorem 2.1.4[EV]. *(a) The formula*

$$\begin{aligned}\{f,g\}_X(p,x,q) =& <p, [\delta_1 f, \delta_1 g]> - <q, [\delta_2 f, \delta_2 g]> \\ & - <\iota\delta_1 f, Dg> - <\iota\delta_2 f, D'g> \\ & + <\iota\delta_1 g, Df> + <\iota\delta_2 g, D'f> \\ & + <R(p)Df, Dg> - <R(q)D'f, D'g>\end{aligned}$$

defines a Poisson bracket on X if and only if $R : U \longrightarrow L(\mathfrak{g}^*, \mathfrak{g})$ is an $ad^*_{\mathfrak{h}}$- equivariant dynamical r-matrix.

(b) The trivial Lie groupoid X equipped with the Poisson bracket $\{\, ,\, \}_X$ in (a) and the Hamiltonian H actions

$$\phi_h^-(p,x,q) = (Ad^*_{h^{-1}}p, hx, q), \quad \phi_h^+(p,x,q) = (p, xh, Ad^*_h q),$$

is a Poisson groupoid in $\mathcal{C}_U$.

The pair $(X, \{\, ,\, \}_X)$ will be called the coboundary dynamical Poisson groupoid associated to R.

Note that the dynamical Poisson groupoid X of Theorem 2.1.4 admits a Poisson groupoid embedding

$$I : H \times U \longrightarrow X : (h,p) \mapsto (Ad^*_{h^{-1}}p, h, p), \tag{2.1.4}$$

where $H \times U$ is the Hamitonian unit. As we will see in later sections, this property turns out to play a crucial role in the study of duality.

We now recall the notion of a Lie algebroid (for more details see [**CdSW**], [**M1**]).

Definition 2.1.5. *A Lie algebroid is a smooth vector bundle $q : A \longrightarrow B$ equipped with a Lie bracket $[\, ,\,]_A$ on the set $\Gamma(A)$ of smooth sections of A and a smooth base preserving bundle map $a : A \longrightarrow TB$, called the anchor map, such that*

$$\begin{aligned} a\,[\zeta, \eta]_A &= [a(\zeta), a(\eta)]_B \\ [\zeta, f\eta]_A &= f\,[\zeta, \eta]_A + a(\zeta)(f)\,\eta,\end{aligned}$$

for all $\zeta, \eta \in \Gamma(A)$ and all $f \in C^\infty(B)$.

The Lie algebroid of a smooth groupoid Γ over B is the vector bundle

$$A(\Gamma) := \big(Ker(T\alpha)\big)_{|\epsilon(B)}$$

over B with anchor map a given by the restriction of $T[\alpha, \beta]$ to $A(\Gamma)$ (here $[\alpha, \beta](z)$ $= (\alpha(z), \beta(z))$, $z \in \Gamma$) and bracket of sections $[X, Y](b) := [X^l, Y^l]_\Gamma(\epsilon(b))$ where

$$X^l : \Gamma \longrightarrow Ker(T\alpha)$$

is the unique left invariant vector field whose restriction to $\epsilon(B)$ is X.

Let V be a vector space and let $\rho : \Gamma \longrightarrow Aut(V)$ be a smooth groupoid morphism where $Aut(V)$ is viewed as a groupoid over its unit element I_V.

Definition 2.1.6. *A smooth map $\Sigma : \Gamma \longrightarrow V$ is called a groupoid 1-cocycle iff*

$$\Sigma(xy) = \Sigma(x) + \rho(x)\Sigma(y)$$

for all $(x, y) \in \Gamma * \Gamma$. The induced map $\Sigma_* : A(\Gamma) \to V$ defined as the restriction of $T\Sigma$ to $A(\Gamma)$ is called the induced Lie algebroid 1-cocycle.

Finally, we recall the notion of an action of a Lie groupoid $\Gamma \rightrightarrows B$ on a manifold S with moment map $f : S \longrightarrow B$. (We follow the terminology of [**MW**].)

Let

$$\Gamma *_f S = \{(x, s) \in \Gamma \times S \mid \beta(x) = f(s)\},$$
$$S *_f \Gamma = \{(s, x) \in S \times \Gamma \mid f(s) = \alpha(x)\}.$$

Definition 2.1.7. *(a) A left action of Γ on S with moment map f is a smooth map $\phi^l : \Gamma *_f S \longrightarrow S : (x, s) \mapsto x \cdot s$ such that*

$$f(x \cdot s) = \alpha(x), \quad y \cdot (x \cdot s) = (yx) \cdot s, \quad \epsilon(f(t)) \cdot t = t,$$

for all $(y, x) \in \Gamma * \Gamma$, $(x, s) \in \Gamma *_f S, t \in S$.

(b) A right action of Γ on S with moment map f is a smooth map $\phi^r : S *_f \Gamma \longrightarrow S : (s, x) \mapsto s \cdot x$ such that

$$f(s \cdot x) = \beta(x), \quad (s \cdot x) \cdot y = s \cdot (xy), \quad t \cdot \epsilon(f(t)) = t,$$

for all $(x, y) \in \Gamma * \Gamma, (s, x) \in S *_f \Gamma, t \in S$.

2.2. Trivial Lie groupoids in $\mathcal{C}_U$

Our purpose in this subsection is to provide an explicit class of Poisson brackets on trivial Lie groupoids which extends the construction of Theorem 2.1.4 (a), and is essential for our subsequent study of duality.

Throughout the paper, we define the bundle map $\Pi^\#$ associated with the Poisson manifold $(Y, \{,\}_Y)$ using the convention $<df, \Pi^\# dg> = \{f, g\}_Y$. Also, we assume that the Lie subgroup $H \subset G$ is connected. We begin with a general property.

Proposition 2.2.1. *Let Y be a Poisson groupoid over U with target and source maps α and β and unit map ϵ. If there exists a (base preserving) Poisson groupoid morphism*

$$I : H \times U \to Y,$$

(here $H \times U$ is the Hamiltonian unit) then Y belongs to $\mathcal{C}_U$.

Proof. From a general property of Poisson groupoids [**W1**], we have

$$\{\alpha^*\varphi, \beta^*\psi\}_Y = 0, \ \forall \varphi, \psi \in C^\infty(U).$$

So it remains to show that Y admits two commuting Hamiltonian H-actions ϕ^-, ϕ^+ with equivariant momentum maps α and β.

Set, as in Definition 2.1.7,

$$(H \times U) *_\alpha Y = \{(h, p, y) \mid \beta_0(h, p) = \alpha(y)\},$$
$$Y *_\beta (H \times U) = \{(y, h, p) \mid \beta(y) = \alpha_0(h, p)\}.$$

Here α_0, β_0 are the target and source maps in Example 2.1.3.

The morphism I induces a left (resp. right) groupoid action of $H \times U$ on Y with moment map α (resp. β), given by:

$$\phi^- : (H \times U) *_\alpha Y \longrightarrow Y$$
$$(h, \alpha(y), y) \mapsto I(h, \alpha(y)) \cdot y$$
$$\phi^+ : Y *_\beta (H \times U) \longrightarrow Y$$
$$(y, h, Ad_h^* \beta(y)) \mapsto y \cdot I(h, Ad_h^* \beta(y))$$

which, upon the natural identifications

$$(H \times U) *_\alpha Y \simeq H \times Y, \ Y *_\beta (H \times U) \simeq Y \times H,$$

induce a left and a right action of H on Y (also denoted by $\phi^\pm$).

We now show that ϕ^- is Hamiltonian with momentum map α (the verification for ϕ^+ and β is similar).

Note that α is equivariant since $\alpha(\phi_k^-(y)) = \alpha(I(k, \alpha(y)) \cdot y) = \alpha(I(k, \alpha(y))) = \alpha_0(k, \alpha(y)) = Ad_{k^{-1}}^* \alpha(y)$.

Let $Z^-(y) = \frac{d}{dt}|_{t=0} \phi_{e^{tZ}}^-(y)$ be the infinitesimal generator of the action corresponding to Z. We want to show Z^- coincides with the Hamiltonian vector field $\widehat{X}_{f_Z \circ \beta}$ where $f_Z \in C^\infty(U)$ is defined by $f_Z(q) = <Z, q>, \forall q \in U$.

Since $I(1, q) = \epsilon(q)$, we have

$$Z^-(y) = \frac{d}{dt}|_{t=0} I(e^{tZ}, \alpha(y)) \cdot y = T_{\epsilon \circ \alpha(y)} r_y T_{(1, \alpha(y))} I(Z, 0).$$

In particular, $Z^- \in Ker T\beta$. On the other hand,

$$Z^-(\epsilon \circ \alpha(y)) = \frac{d}{dt}|_{t=0} I(e^{tZ}, \alpha(y)) \cdot I(1, \alpha(y))$$
$$= \frac{d}{dt}|_{t=0} I(e^{tZ}, \alpha(y))$$
$$= T_{(1, \alpha(y))} I(Z, 0),$$

thus Z^- is right invariant. Since $\widehat{X}_{f_Z \circ \alpha}$ is also right invariant [**X**], it suffices to show that both vector fields coincide on $\epsilon(U)$. But from the Poisson property of I,

we have
$$\widehat{X}_{f_Z \circ \alpha}(\epsilon \circ \alpha(y)) = \Pi_Y^{\#}(\epsilon \circ \alpha(y))d(f_Z \circ \alpha)$$
$$= T_{(1,\alpha(y))}I\,\Pi_0^{\#}(1,\alpha(y))d(f_Z \circ \alpha \circ I)$$
$$= T_{(1,\alpha(y))}I\,\Pi_0^{\#}(1,\alpha(y))d(f_Z \circ \alpha_0)$$
$$= T_{(1,\alpha(y))}I\,\Pi_0^{\#}(1,\alpha(y))(ad_Z^*\alpha(y), Z)$$
$$= T_{(1,\alpha(y))}I\,(Z,0).$$

Now, since Z^- is Hamiltonian, its flow $\phi_{\epsilon Z}^-$ at $t=1$ preserves the Poisson bracket of Y. Therefore, the connectedness of H implies that ϕ^- is Hamiltonian with momentum map α. Hence the claim. ∎

For the rest of this subsection, we let $X = U \times G \times U$ be the trivial Lie groupoid of section 2.1 (see (2.1.3)). We will describe all pairs
$$(X, \{\,,\,\}), \quad I : H \times U \to X,$$
where $\{\,,\,\}$ is a Poisson bracket on X compatible with its groupoid structure and I is a morphism of Poisson groupoids, where $H \times U$ is the Hamiltonian unit.

Let $\rho : G \longrightarrow Aut(V)$ be a representation of G on the vector space V. We are going to restrict ourselves to groupoid 1 cocycles $P : X \longrightarrow V$ which satisfy
$$P(p, xy, q) = P(p, x, r) + \rho(x)P(r, y, q) \text{ for all } p, q, r \in U, x, y \in G.$$

Proposition 2.2.2. *P is a 1-cocycle on X iff*
$$P(p, x, q) = -l(p) + \pi(x) + \rho(x)l(q),$$
where $l : U \longrightarrow V$ is a smooth map with $l(q_0) = 0$ for some $q_0 \in U$, and $\pi : G \longrightarrow V$ is a group 1-cocycle.

Proof. Clearly any such map is a 1-cocycle. Conversely, if P is a cocycle then $P(p, 1, p) = P(p, 1, q_0) + P(q_0, 1, p) = 0$ and $\pi(x) = P(q_0, x, q_0)$ is a group cocycle. The claim then follows from $P(p, x, q) = P(p, x, q_0) + \rho(x)P(q_0, 1, q) = P(p, 1, q_0) + P(q_0, x, q_0) + \rho(x)P(q_0, 1, q) = -P(q_0, 1, p) + \pi(x) + \rho(x)P(q_0, 1, q).$ ∎

Proposition 2.2.3. *Let $\Pi \in \Gamma(\bigwedge^2 TX)$ be a bivector field. Then the graph of $m : X * X \longrightarrow X$ is Π-coisotropic in $X \times X \times \overline{X}$ iff*
$$\Pi^{\#}(p, x, q)(Z_1, B, Z_2) = (-K(p)Z_1 - A^*(p)T_1^*r_x B,$$
$$T_1 r_x A(p)Z_1 + T_1 l_x A(q)Z_2 + T_1 r_x P(p, x, q)T_1^*r_x B,$$
$$K(q)Z_2 - A^*(q)T_1^*l_x B),$$

where $K : U \longrightarrow L(\mathfrak{h}, \mathfrak{h}^*)$ and $A : U \longrightarrow L(\mathfrak{h}, \mathfrak{g})$ are smooth maps, and $P : X \longrightarrow L(\mathfrak{g}^*, \mathfrak{g})$ is a groupoid 1-cocycle for the adjoint action. Here, K and P are pointwise skew-symmetric.

Proof. The graph of the multiplication m is
$$Gr(m) = \{\big((p, x, q), (q, y, r), (p, xy, r)\big)\} \subset X \times X \times \overline{X}.$$

Therefore, $\Omega \in \left(T_{((p,x,q),(q,y,r),(p,xy,r))} Gr(m)\right)^{\perp}$ if and only if

$$\Omega = ((Z_1, \omega, Z_2), (-Z_2, T_y^*(r_{y^{-1}} \circ l_x)\omega, Z_3), (-Z_1, -T_{xy}^* r_{y^{-1}}\omega, -Z_3)),$$

for some $Z_1, Z_2, Z_3 \in \mathfrak{h}$ and $\omega \in (T_x G)^*$.

One then verifies (see the appendix for details) that the Π-coisotropy of $Gr(m)$:

$$(\Pi \oplus \Pi \oplus -\Pi)(\Omega, \Omega') = 0, \quad \forall \Omega, \Omega' \in (TGr(m))^{\perp},$$

is equivalent to our assertion. ∎

Now, a map $I : H \times U \longrightarrow X$ is a (base preserving) groupoid morphism iff

$$I(k,q) = (Ad^*_{k^{-1}} q, \chi(k,q), q) \tag{2.2.1}$$

for some smooth map χ satisfying

$$\chi(hk, q) = \chi(h, Ad^*_{k^{-1}} q)\chi(k,q). \tag{2.2.2}$$

In particular, $\chi(1,q) = 1$, and if $0 \in U$, the map $H \longrightarrow G : k \mapsto \chi(k, 0)$ is a group morphism.

Note that (2.2.2) implies that $\chi : H \times U \longrightarrow G$ is a groupoid morphism when G is viewed as a groupoid over its unit element. Applying the Lie functor to χ then provides an algebroid morphism

$$A(\chi) : U \times \mathfrak{h} \longrightarrow \mathfrak{g}$$
$$(q, Z) \mapsto T_{(1,q)} \chi(Z, ad^*_Z q),$$

which we will henceforth denote as $(q, Z) \mapsto A_\chi(q) Z$. The morphism property then implies that for all $Z, Z' \in \mathfrak{h}, p \in U$,

$$A_\chi(p)[Z, Z'] = dA_\chi(p) \cdot ad^*_Z p \cdot Z' - dA_\chi(p) \cdot ad^*_{Z'} p \cdot Z \tag{2.2.3}$$
$$+ [A_\chi(p) Z, A_\chi(p) Z'].$$

Proposition 2.2.4. *If (X, Π) is a Poisson groupoid with $\Pi^\#$ expressed as in Proposition 2.2.3 above, then the map I is a Poisson map iff*

(a) $\quad K(p) Z = ad^*_Z p, \quad \forall Z, Z' \in \mathfrak{h}.$

(b) $\quad A_\chi(p) = A(p)$

(c) *For all $\alpha, \beta \in \mathfrak{g}^*$,*

$$<\alpha, P(I(h, p))\beta> = <\lambda_\alpha, Z_\beta> - <\lambda_\beta, Z_\alpha> - <p, [Z_\alpha, Z_\beta]>,$$

where $\lambda_\alpha \in \mathfrak{h}^*$ and $Z_\alpha \in \mathfrak{h}$ are defined by

$$<\lambda_\alpha, Z> = <\alpha, T_{(h,p)}(r_{(\chi(h,p))^{-1}} \circ \chi)(T_1 l_h Z, 0)>, \quad \forall Z \in \mathfrak{h}$$
$$<Z_\alpha, \lambda> = <\alpha, T_{(h,p)}(r_{(\chi(h,p))^{-1}} \circ \chi)(0, \lambda)>, \quad \forall \lambda \in \mathfrak{h}^*.$$

Proof. We have to impose the condition

$$\{f \circ I, g \circ I\}_{H \times U} = \{f, g\} \circ I,$$

where $\{,\}_{H\times U}$ is the bracket in Example 2.1.3. Let $i_h : U \to H \times U : p \mapsto (h,p)$ and $i_p : H \to H \times U : h \mapsto (h,p)$. We have

$$\delta(f \circ I)(h,p) = Ad_{h^{-1}}\delta_1 f + T_p^*(\chi \circ i_h)\partial f + \delta_2 f$$
$$D'(f \circ I)(h,p) = ad^*_{Ad_{h^{-1}}\delta_1 f} p + T_1^*(\chi \circ i_p \circ l_h)\partial f.$$

Here (and below) we will use the shorthand notation d_{ij}, $i,j \in \{1,2,*\}$ to stand for $\{p_i^*\phi \circ I, p_j^*\psi \circ I\}_{H\times U} - \{p_i^*\phi, p_j^*\psi\} \circ I$, where, as an index, $* = G$ and p_1, p_G, p_2 are the projections onto the first, second, and third factor of X respectively. Thus for example

$$d_{1*} = \{p_1^*\phi \circ I, p_G^*\psi \circ I\}_{H\times U} - \{p_1^*\phi, p_G^*\psi\} \circ I.$$

It is easy to see that $d_{12} = 0$, and that $d_{11} = 0 \Leftrightarrow d_{22} = 0 \Leftrightarrow (a)$ holds. Now

$$d_{1*} = d\psi\big(-T_1(\chi \circ i_p \circ r_h)\delta\phi + T_1 r_{\chi(h,p)} A(Ad_{h^{-1}}^* p)\delta\phi\big)$$
$$= d\psi\big(-\frac{d}{dt}\Big|_0 \chi(e^{t\delta\phi}h, p) + T_1 r_{\chi(h,p)} A(Ad_{h^{-1}}^* p)\delta\phi\big)$$
$$= d\psi\, T_1 r_{\chi(h,p)}\big(-\frac{d}{dt}\Big|_{t=0} \chi(e^{t\delta\phi}, Ad_{h^{-1}}^* p) + A(Ad_{h^{-1}}^* p)\delta\phi\big)$$

(by (2.2.2))

$$= d\psi T_1 r_{\chi(h,p)}\big(-A_\chi(Ad_{h^{-1}}^* p) + A(Ad_{h^{-1}}^* p)\big)\delta\phi,$$

where in the last two steps we have used $\chi(1,q) = 0, \forall q \in U$. Thus $d_{1*} = 0 \Leftrightarrow (b)$ holds. Similarly one has $d_{2*} = 0 \Leftrightarrow (b)$ holds. Finally,

$$d_{**} = <\lambda_{D\psi}, Z_{D\psi'}> - <\lambda_{D\psi'}, Z_{D\psi}> - <p, [Z_{D\psi}, Z_{D\psi'}]>$$
$$- <D\psi, P \circ I(h,p) D\psi'>.$$

Hence the claim. ∎

Assembling the above propositions, we can now formulate the main assertion of this subsection.

Theorem 2.2.5. *If $(X, \{,\})$ is a Poisson groupoid which admits a Poisson groupoid morphism $I : H \times U \to X$ as in (2.2.1). Then*

(a) The Poisson bracket must be of the form

$$\{f,g\}(p,x,q) = <p, [\delta_1 f, \delta_1 g]> - <q, [\delta_2 f, \delta_2 g]>$$
$$- <A_\chi(p)\delta_1 f, Dg> - <A_\chi(q)\delta_2 f, D'g>$$
$$+ <A_\chi(p)\delta_1 g, Df> + <A_\chi(q)\delta_2 g, D'f>$$
$$+ <Df, P(p,x,q)Dg>,$$

where the groupoid 1-cocycle P satisfies Proposition 2.2.4 (c).

(b) The Jacobi identity for $\{,\}$ is equivalent to the condition

$$<\beta, [P\alpha, P\gamma]> - <\beta, DP \cdot P\alpha \cdot \gamma> + <\beta, \delta_1 P \cdot (A_\chi(p)^*\alpha) \cdot \gamma>$$
$$+ <\beta, \delta_2 P \cdot (A_\chi(q)^* Ad_x^* \alpha) \cdot \gamma> + c.p.(\alpha, \beta, \gamma) = 0$$

for all $\alpha, \beta, \gamma \in \mathfrak{g}^*$, where P stands for $P(p, x, q)$ and

$$\delta_1 P(\lambda) = \frac{d}{dt}_{|t=0} P(p + t\lambda, x, q), \quad \delta_2 P(\lambda) = \frac{d}{dt}_{|t=0} P(p, x, q + t\lambda),$$

$$DP \cdot X = \frac{d}{dt}_{|t=0} P(p, e^{tX} x, q).$$

(c) $(X, \{,\})$ belongs to $\mathcal{C}_U$ with Hamiltonian H- actions

$$\phi^- : H \times X \longrightarrow X$$
$$(h, p, x, q) \mapsto (Ad^*_{h^{-1}} p, \chi(h, p)x, q)$$
$$\phi^+ : X \times H \longrightarrow X$$
$$(p, x, q, h) \mapsto (p, x\chi(h, Ad^*_h q), Ad^*_h q).$$

Proof. (a) This assertion is simply a restatement of Proposition 2.2.3 and Proposition 2.2.4.

(b) We give the main steps (see the appendix for the details of the calculations):

For the Jacobi identity, we use the shorthand notation J_{ijk}, $i, j, k \in \{1, 2, *\}$ to stand for $\{p_i^* \varphi, \{p_j^* \varphi', p_k^* \psi\}\} + c.p.$, where, as an index, $* = G$, and p_1, p_G, p_2 are as in the proof of Proposition 2.2.4. Thus for example $J_{12*} = \{p_1^* \varphi, \{p_2^* \varphi', p_G^* \psi\}\} + c.p.$.

Clearly, we have $J_{ijk} = 0$ for $i, j, k \in \{1, 2\}$ and $J_{*12} = 0$ and these do not impose any conditions. On the other hand, $J_{*11} = 0 \Leftrightarrow J_{*22} = 0 \Leftrightarrow A_\chi$ satisfies (2.2.3). Writing $P(p, x, q) = -l(p) + \pi(x) + Ad_x^* l(q) Ad_x^*$ as in Proposition 2.2.2, we have

$$J_{**1} = 0 \Leftrightarrow J_{**2} = 0 \Leftrightarrow$$
$$< \alpha, (dl(p) ad_Z^* p + ad_{A_\chi(p)Z} l(p) + l(p) ad^*_{A_\chi(p)Z} + d\pi(1) A_\chi(p) Z) \beta > \quad (\star)$$
$$= + < dA_\chi(p)(A_\chi(p)^* \alpha) Z, \beta > - < dA_\chi(p)(A_\chi(p)^* \beta) Z, \alpha >, \quad \forall \alpha, \beta \in \mathfrak{g}^*.$$

But the latter follows upon differentiating the identity of Proposition 2.2.4 (c) at $(1, p)$. Indeed, that $lhs(\star) = \frac{d}{dt}_{|t=0} < \alpha, P \circ I(e^{tZ}, p)\beta >$ is clear. On the other hand, upon using

$$< \lambda, Z_\alpha(1, p) > = < \alpha, \frac{d}{dt}_{|t=0} \chi(1, p + t\lambda) \chi(1, p)^{-1} > = 0$$

$$< \lambda_\alpha(1, p), Z > = < \alpha, \frac{d}{dt}_{|t=0} \chi(e^{tZ}, p) \chi(1, p)^{-1} > = < A_\chi(p)^* \alpha, Z >,$$

and

$$< \lambda, \frac{d}{dt}_{|t=0} Z_\beta(e^{tZ}, p) > = < \beta, \frac{d}{dt}_{|t=0} (\frac{d}{ds}_{|s=0} (\chi(e^{tZ}, p + s\lambda) \chi(e^{tZ}, p)^{-1})) >$$
$$= < \beta, dA_\chi(p)(\lambda) Z >,$$

we have

$$\frac{d}{dt}_{|t=0}\Big(<\lambda_\alpha,Z_\beta>(e^{tZ},p)-<\lambda_\beta,Z_\alpha>(e^{tZ},p)-<p,[Z_\alpha(e^{tZ},p),Z_\beta(e^{tZ},p)]>\Big)$$

$$=<A_\chi(p)^*\alpha,\frac{d}{dt}_{|t=0}Z_\beta(e^{tZ},p)>-<A_\chi(p)^*\beta,\frac{d}{dt}_{|t=0}Z_\alpha(e^{tZ},p)>$$

$$=rhs(\star).$$

Thus, with our assumptions, the bracket $\{\,,\,\}$ satisfies the Jacobi identity if and only if $J_{***}=0$ which is precisely (b).

(c) This is Proposition 2.2.1: The morphism $I:H\times U\to X$ induces a left groupoid action of $H\times U$ on X over α_X

$$\phi^-:(H\times U)*_{\alpha_X}X\longrightarrow X:((h,p),(p,x,q))\mapsto I(h,p)\cdot(p,x,q)$$

and a right groupoid action on X over β_X

$$\phi^+:X*_{\beta_X}(H\times U)\longrightarrow X:((p,x,q),(k,Ad_k^*q))\mapsto(p,x,q)\cdot I(k,Ad_k^*q).$$

With the identifications $(H\times U)*_{\alpha_X}X\simeq H\times X$ and $X*_{\beta_X}(H\times U)\simeq X\times H$, these actions are the ones given above. ∎.

We end this subsection with a definition and some remarks.

Definition 2.2.6. *Following Etingof and Varchenko, the Poisson groupoid $(X,\{\,,\,\})$ of Theorem 2.2.5 is said to be of **dynamical** type iff $I(k,q)=(Ad_{k^{-1}}^*q,k,q)$. In this case, the corresponding Lie algebroid dual $A(X)^*$ (with its natural Lie algebroid structure) will be called a dynamical Lie algebroid.*

Remarks 2.2.7. (a) For X of dynamical type, $\chi(h,p)=\chi(h,1)=h$, thus $A_\chi(p)Z=Z$. Therefore, the first six terms of the Poisson bracket coincide with those of the coboundary case. The last term however, which is given by

$$P(p,x,q)=-l(p)+\pi(x)+Ad_x l(q)Ad_x^*$$

differs from the coboundary case by the group 1-cocycle $\pi:G\to L(\mathfrak{g}^*,\mathfrak{g})$.

As we will demonstrate in section 4 for a class of solutions of the modified dynamical Yang-Baxter equation on simple Lie algebras, Poisson groupoids of dynamical type with $\pi\neq 0$ arise naturally as Poisson groupoid duals (modulo Poisson groupoid isomorphisms) of certain associated coboundary dynamical groupoids. Note that the situation here is analogous to that for Poisson Lie groups: Typically, the Poisson Lie group dual of a Lie group equipped with the Sklyanin bracket is not of coboundary type.

(b) For X of dynamical type, we have

$$<Z_\alpha,\lambda>(h,p)=<\alpha,\frac{d}{dt}_{|t=0}(\chi(h,1)\chi(h,1)^{-1})>=0.$$

Therefore, the identity of Proposition 2.2.4 (c) simplifies to

$$P(Ad_{h^{-1}}^*p,h,p)=0.$$

In other words P vanishes on the H-orbit of $\epsilon(U) \subset X$. This condition is the natural extension of the $\mathfrak{h}$-equivariance of the dynamical r-matrix which it reduces to when $\pi \equiv 0$.

(c) We now comment on the effect of automorphisms of the trivial Lie groupoid $X = U \times G \times U$ on the Poisson bracket of Theorem 2.2.5 as well as on the dynamical property 2.2.6.

Let $(X, \{,\})$ be as in Theorem 2.2.5 and let
$$\Sigma : X \longrightarrow X$$
$$(p, x, q) \mapsto (p, \sigma(p)\, \Phi(x)\, \sigma(q)^{-1}, q)$$
be the (base preserving) groupoid automorphism associated with the smooth map $\sigma : U \longrightarrow G$ and the group automorphism $\Phi \in Aut(G)$. Denote the map
$$T_p(r_{\sigma(p)^{-1}} \circ \sigma) : \mathfrak{h}^* \simeq T_p U \longrightarrow \mathfrak{g}$$
by $D_\sigma(p)$. Then the Poisson bracket of the transported pair
$$(X, \{f, g\}_\Sigma = \{f \circ \Sigma, g \circ \Sigma\} \circ \Sigma^{-1}), \quad I_\Sigma = \Sigma \circ I,$$
is given by Theorem 2.2.5 with
$$A_{\chi_\Sigma}(p)(Z) = Ad_{\sigma(p)}\, T_1\Phi\, A_\chi(p)(Z) - D_\sigma(p)\, ad^*_Z p, \quad Z \in \mathfrak{h},$$
and groupoid 1-cocycle
$$P_\Sigma(p, x, q) = -l_\Sigma(p) + \pi_\Sigma(x) + Ad_x\, l_\Sigma(q)\, Ad^*_x,$$
where
$$l_\Sigma(p)(\alpha) = D_\sigma(p)\, ad^*_{D^*_\sigma(p)\alpha} p$$
$$+ D_\sigma(p) A_\chi(p)^* T_1\Phi^* Ad^*_{\sigma(p)} \alpha - Ad_{\sigma(p)}\, T_1\Phi\, A_\chi(p) D^*_\sigma(p) \alpha$$
$$+ Ad_{\sigma(p)}\, T_1\Phi\, l(p)\, T_1\Phi^*\, Ad^*_{\sigma(p)} \alpha + T_1\Phi\, \pi(\Phi^{-1}(\sigma(p)))\, T_1\Phi^* \alpha,$$

$$\pi_\Sigma(x) = T_1\Phi\, \pi(\Phi^{-1}(x))\, T_1\Phi^*.$$

In particular, it is important to observe that the dynamical property 2.2.6 is not preserved by arbitrary automorphisms of X. Indeed, for X of dynamical type, the transported pair $((X, \{,\}_\Sigma), I_\Sigma)$ is dynamical (for the same inclusion of H into G) if and only if the map σ is twisted H-equivariant:
$$\sigma(Ad^*_{h^{-1}} p) = h\, \sigma(p)\, \Phi(h)^{-1}, \quad \forall h \in H.$$

Note that, for X dynamical, when $\pi \equiv 0$ and σ is twisted H-equivariant, the map $l \mapsto l_\Sigma$ induces a group action of $C^\infty(U, G)^H$ on the set of dynamical r-matrices. This action in the special instance $\Phi = Id_G$ has played a key role in the classification theorem of Etingof and Schiffmann in [ES].

(d) As a special instance of Theorem 2.2.5 (see also Example 3.1.2), consider $X = \mathfrak{h}^* \times G \times \mathfrak{h}^*$ equipped with the Poisson bracket
$$\{f, g\}(p, x, q) = <p, [\delta_1 f, \delta_1 g]> - <q, [\delta_2 f, \delta_2 g]>,$$
and morphism
$$I : H \times \mathfrak{h}^* \longrightarrow X : (h, p) \mapsto (Ad^*_{h^{-1}} p, 1, p).$$

It is immediate from Remark (c) above that X is not isomorphic to a Poisson groupoid of dynamical type.

CHAPTER 3

Duality

3.1. Duality of Poisson groupoids

Following [**W1**],[**M2**], [**MX2**], we begin by recalling the notion of duality of Poisson groupoids and the definition of the dual (when it exists) of a Poisson groupoid.

Let $(Y, \{,\}_Y)$ be a Poisson groupoid over B with target and source maps α, β, and unit map ϵ. We define, as in section 2, the bundle map $\Pi^\#$ by the formula $\{f, g\}_Y = <df, \Pi^\# dg>$.

Since Y is Poisson, the set of 1-forms $\Omega^1(Y)$ inherits a Lie bracket from $C^\infty(Y)$ [**Ko**], [**MM**],[**W1**], given by

$$[\omega, \omega'] = -L_{\Pi^\#\omega}\omega' + L_{\Pi^\#\omega'}\omega - d<\omega, \Pi^\#\omega'>, \qquad (3.1.1)$$

and the map

$$\Omega^1(Y) \longrightarrow \mathfrak{X}(Y) : \omega \mapsto -\Pi^\#\omega$$

is a morphism of Lie algebras, where $\mathfrak{X}(Y)$ is the set of vector fields on Y with the usual Lie bracket. Therefore, T^*Y is a Lie algebroid over Y.

Now, it follows from a general result in [**W1**] that the unit submanifold $\epsilon(B)$ of the Poisson groupoid Y is coisotropic in Y, hence its conormal bundle

$$N^*\big(\epsilon(B)\big) := \bigcup_{q \in B} \big(T_{\epsilon(q)}\epsilon(B)\big)^\perp \subset T^*Y_{|\epsilon(B)}$$

inherits a Lie algebroid structure: the bracket of two sections $\theta_1, \theta_2 : B \longrightarrow N^*(\epsilon(B))$ is

$$[\theta_1, \theta_2]_{N^*}(q) = [\bar\theta_1, \bar\theta_2](\epsilon(q)) \qquad (3.1.2)$$

for arbitrary $\bar\theta_1, \bar\theta_2 \in Ker\, \epsilon^*$ subject to $\bar\theta_1 \circ \epsilon = \theta_1, \bar\theta_2 \circ \epsilon = \theta_2$, while the anchor map $a_* : N^*\big(\epsilon(B)\big) \to TB$ is given by the restriction of $-\Pi^\#$ to $N^*\big(\epsilon(B)\big)$.

Since we have a natural identification $N^*\big(\epsilon(B)\big) \simeq A(Y)^*$, we will therefore always take $A(Y)^*$ with the induced Lie algebroid structure and the pair $(A(Y), A(Y)^*)$ will be called the tangent Lie bialgebroid of $(Y, \{,\})$. For the precise definition of Lie bialgebroids see [MX1]; note however that (A, A^*) is a Lie bialgebroid if and only if (A^*, A) is.

Definition 3.1.1 [**M2**]. *Two Poisson groupoids Y and Y' over the same base are in duality if and only if the Lie bialgebroids $(A(Y), A(Y)^*)$ and $(A(Y')^*, A(Y')^-)$ are isomorphic. Here, $A(Y')^-$ is obtained from $A(Y')$ by changing the sign of both anchor and bracket of sections.*

Note that if the Lie algebroid $A(Y)^*$ is integrable, then there exists (by Lie I) a unique source-simply connected Lie groupoid Y^* integrating $A(Y)^*$. In this case, it follows from a general theorem of Mackenzie and Xu [**MX2**] that the latter may be equipped with a unique Poisson bracket $\{\,,\,\}_{Y^*}$ compatible with its groupoid structure such that $(Y^*, \{\,,\,\}_{Y^*})$ has tangent Lie bialgebroid $(A(Y)^*, A(Y))$.

The Poisson groupoids Y and Y^* are Poisson groupoids in duality and $(Y^*, \{\,,\,\}_{Y^*})$ is called the dual of $(Y, \{\,,\,\}_Y)$.

The following example is already in [**W1**]:

Example 3.1.2. The Poisson groupoid dual to the Hamiltonian unit $H \times U$ of Example 2.1.3 is the coarse groupoid $U \times U$ with Poisson bracket

$$\{f, g\}_{U \times U}(p, q) = <p, [\delta_1 f, \delta_1 g]> - <q, [\delta_2 f, \delta_2 g]>.$$

Note that $U \times U$ belongs to $\mathcal{C}_U$ with the Hamiltonian H-actions

$$\phi_k^-(p, q) = (Ad^*_{k^{-1}} p, q), \quad \phi_k^+(p, q) = (p, Ad^*_k q).$$

The associated tangent Lie bialgebroid is given by

$$\big(A(H \times U), A(U \times U)\big) = \big(U \times \mathfrak{h}, U \times \mathfrak{h}^*\big)$$

and the respective Lie brackets on smooth sections are as follows :

$$[Z_1, Z_2](q) = [Z_1(q), Z_2(q)]_\mathfrak{h} + dZ_2(q) ad^*_{Z_1(q)} q - dZ_1(q) ad^*_{Z_2(q)} q, \quad Z_1, Z_2 : U \to \mathfrak{h}$$

$$[X_1, X_2](q) = dX_2(q) X_1(q) - dX_1(q) X_2(q), \quad X_1, X_2 : U \to \mathfrak{h}^*.$$

We now recall a special (and simplest) instance of Lie bialgebroid morphisms. Let (A, A^*) and $(A', (A')^*)$ be two Lie bialgebroids over B with bundle projections $q : A \to B$, $q_* : A^* \to B$, anchors $a : A \to TB$, $a_* : A^* \to TB$, and similarly for $(A', (A')^*)$.

Definition 3.1.3. *A bundle map*

$$\phi : A \longrightarrow A'$$
$$q \searrow \quad \swarrow q'$$
$$B$$

is called a Lie bialgebroid morphism if and only if both ϕ and ϕ^* are Lie algebroid morphisms, i.e.

$$a' \circ \phi = a, \quad \phi[X, Y]_A = [\phi(X), \phi(Y)]_{A'}$$
$$a_* \circ \phi^* = a'_*, \quad \phi^*[\alpha', \beta']_{(A')^*} = [\phi^*(\alpha'), \phi^*(\beta')]_{A^*}$$
for all sections $X, Y : B \to A$, $\alpha', \beta' : B \to (A')^*$.

The following property [**MX1**] is basic.

Proposition 3.1.4. *Let Y, Y' be Poisson groupoids over B with tangent Lie bialgebroids (A, A^*) and $(A', (A')^*)$. If $\mu : Y \longrightarrow Y'$ is a base preserving morphism of Poisson groupoids, then*

$$A(\mu) : A \longrightarrow A'$$

is a (base preserving) morphism of Lie bialgebroids.

3.2. The dual of a dynamical Poisson groupoid

Throughout this subsection, we will equip $X = U \times G \times U$ with the trivial Lie groupoid structure and we assume that the subgroup $H \subset G$ is connected and simply connected. We begin with a description of the tangent Lie bialgebroid of a Poisson groupoid $(X, \{\,,\,\})$, a special instance of which is described in [**BKS**].

Proposition 3.2.1. *Let $(X, \{\,,\,\})$ be a Poisson groupoid with Poisson bracket as in Proposition 2.2.3. Then the Lie bialgebroid tangent to X is (isomorphic to) the pair $\left(U \times \mathfrak{h}^* \times \mathfrak{g},\, U \times \mathfrak{h} \times \mathfrak{g}^*\right)$ with anchor maps*

$$a : U \times \mathfrak{h}^* \times \mathfrak{g} \to U \times \mathfrak{h}^* : (q, \lambda, X) \mapsto (q, \lambda)$$
$$a_* : U \times \mathfrak{h} \times \mathfrak{g}^* \to U \times \mathfrak{h}^* : (q, Z, B) \mapsto (q, -K(q)Z + A^*(q)B)$$

and Lie brackets of sections $[\,,\,]$, $[\,,\,]_*$ given by the following expressions

$$[(\lambda, X), (\lambda', X')](q) = (d\lambda' \cdot \lambda - d\lambda \cdot \lambda',\, dX' \cdot \lambda - dX \cdot \lambda' + [X(q), X'(q)]_\mathfrak{g})$$

$$< [(Z, B), (Z', B')]_*, (\Lambda, Y) > (q) =$$
$$< -dZ'(K(q)Z - A^*(q)B) + dZ(K(q)Z' - A^*(q)B'), \Lambda >$$
$$- < Z', dK(q)(\Lambda)Z >$$
$$- < B, \delta_1 P(\Lambda) B' > - < B', dA(q)(\Lambda)\,Z > + < B, dA(q)(\Lambda) Z' >$$
$$+ < dB(K(q)Z' - A^*(q)B') - dB'(K(q)Z - A^*(q)B), Y >$$
$$+ < ad^*_{A(q)Z} B' - ad^*_{A(q)Z'} B, Y >$$
$$+ < B, \partial P(Y) B' >,$$

where $\lambda, \lambda' : U \to \mathfrak{h}^*; X, X' : U \to \mathfrak{g}$; $Z, Z' : U \to \mathfrak{h}; B, B' : U \to \mathfrak{g}^*$ are smooth maps, $\Lambda \in \mathfrak{h}^*, Y \in \mathfrak{g}$, all differentials and sections are evaluated at q, and the partial derivatives of the groupoid 1-cocycle P (see Theorem 2.2.5) are evaluated at $(q, 1, q)$.

Proof. The Lie algebroid $A(X)$ is well known (see, for example, [**M1**]). Although for the coboundary dynamical case an algebraic description of the Lie algebroid dual $A(X)^*$ was given in [**BKS**], it was not derived there from the Poisson groupoid using Weinstein's coisotropic calculus. So we will briefly indicate the steps of the calculation.

The unit section of X is given by $\epsilon : U \to U \times G \times U : q \mapsto (q, 1, q)$. Therefore, $\gamma \in N(\epsilon(U))_q$ if and only if $\gamma = (-Z, B, Z)$, for some $Z \in \mathfrak{h}$ and $B \in \mathfrak{g}^*$.

Let $\alpha, \alpha' : U \to N(\epsilon(U))$ be two sections written as $\alpha(q) = (-Z(q), B(q), Z(q))$ and $\alpha'(q) = (-Z'(q), B'(q), Z'(q))$. Set $\omega(p, x, q) = (-Z(q), T_x^* l_{x^{-1}} B(q), Z(q))$, and similarly for ω'. By (3.1.2), it suffices

1) to compute
$$[\alpha, \alpha']_{N(\epsilon(U))}(q) = [\omega, \omega'](q, 1, q),$$

where the right hand side is given by (3.1.1) with Hamiltonian operator

$$\Pi^\#(p,x,q)(Z_1, B, Z_2) = (-K(p)Z_1 - A^*(p)T_1^* r_x B,$$
$$T_1 r_x A(p) Z_1 + T_1 l_x A(q) Z_2 + T_1 r_x P(p,x,q) T_1^* r_x B,$$
$$K(q) Z_2 - A^*(q) T_1^* l_x B),$$

and

2) to choose an identification of $N(\epsilon(U))$ with $U \times \mathfrak{h} \times \mathfrak{g}^*$.

The computation for (1) is rather lengthy (for a brief sketch see the appendix) and may be performed with the help of

$$< L_{\Pi^\# \omega} \omega', (\Lambda, X^l, \Lambda') >$$
$$= < d < \omega', (\Lambda, X^l, \Lambda') >, \Pi^\# \omega > + < \omega', [(\Lambda, X^l, \Lambda'), \Pi^\# \omega] >,$$

where $\Lambda, \Lambda' \in \mathfrak{h}^*$, $X \in \mathfrak{g}$, and X^l is the left invariant vector field on G with $X^l(1) = X$.

As for 2) the natural identification to make is given by $\iota_- : N(\epsilon(U))_q \to \mathfrak{h} \times \mathfrak{g}^* : (-Z, B, Z) \mapsto (Z, B)$. Setting $[(Z, B), (Z', B')]_*(q) = \iota_-[\omega, \omega'](q, 1, q)$ then gives the stated formula. ∎

Remarks 3.2.2. (a) The coboundary dynamical case in [**BKS**] corresponds to the choice: $\chi(h, q) = h$, and the groupoid 1-cocycle

$$P(p, x, q) = -R(p) + Ad_x R(q) Ad_x^*.$$

Note that $A(q) Z = A_\chi(q) Z = \iota Z$ is constant in this case, while the induced algebroid 1− cocycle

$$P_* : U \times \mathfrak{h}^* \times \mathfrak{g} \longrightarrow L(\mathfrak{g}^*, \mathfrak{g})$$
$$(q, \Lambda, Y) \mapsto (\partial P \cdot Y + \delta_2 P \cdot \Lambda)(q, 1, q) = (\partial P \cdot Y - \delta_1 P \cdot \Lambda)(q, 1, q)$$

which appears in the bracket of Proposition 3.2.1 is given by

$$P_*(q, \Lambda, Y) = dR(q)\Lambda + R(q) ad_Y^* + ad_Y R(q).$$

(b) Note that Lie bialgebroid structures on (A, A^*) where A is a transitive Lie algebroid $TM \oplus (M \times \mathfrak{g})$ with $H^1(M) = 0$ have recently been classified in [**LiX**].

The proposition which follows is a special instance of the functorial relationship between Poisson groupoids and Lie bialgebroids.

Proposition 3.2.3. Let $(X, \{ , \})$ be a Poisson groupoid with Poisson bracket as in Proposition 2.2.3 and let

$$I : H \times U \longrightarrow U \times G \times U$$
$$(h, p) \mapsto (Ad_{h^{-1}}^* p, \chi(h, p), p)$$

be a groupoid morphism. If $A(I)$ is a morphism of Lie bialgebroids, then I is also a Poisson map. Hence I is a Poisson groupoid morphism.

Proof. We have to show that the three conditions of Propostion 2.2.4 are satisfied. From the definition of I, it is clear that the induced morphism $A(I) : U \times \mathfrak{h} \longrightarrow U \times \mathfrak{g} \times \mathfrak{h}^*$ is given by

$$(q, Z) \mapsto (q, A_\chi(q)(Z), ad_Z^* q),$$

so its dual map $A(I)^* : U \times g^* \times \mathfrak{h} \longrightarrow U \times \mathfrak{h}^*$ is of the form

$$(q, B, Z) \mapsto (q, A_\chi^*(q)B - ad_Z^* q),$$

where $A_\chi(q)(Z) = A(\chi)(q, Z)$. Thus $A(I)^*$ preserves the anchor maps if and only if

$$K(q)(Z) = ad_Z^* q \text{ and } A^*(q) = A_\chi^*(q).$$

We now impose the condition that $A(I)^*$ be a Lie algebroid morphism. Let

$$d((Z, B), (Z', B')) := A(I)^*[(Z, B), (Z', B')]_* - [A(I)^*(Z, B), A(I)^*(Z', B')]_{U \times \mathfrak{h}^*},$$

where $[\,,\,]_*$ is given by Proposition 3.2.1, and $[\,,\,]_{U \times \mathfrak{h}^*}$ is as in Example 3.1.2.

A direct calculation shows that $d((Z, 0), (Z', 0)) = 0$, while

$$d((Z, 0), (0, B')) = 0 \Leftrightarrow A_\chi \text{ satisfies (2.2.3)}.$$

On the other hand,

$$d((0, B), (0, B')) = 0 \Leftrightarrow$$
$$< B, \partial P(A_\chi(\mathbf{Z}))B' - \delta_1 P(ad_{\mathbf{Z}}^* q)B' >$$
$$=< B', dA_\chi(A_\chi^*(B))\mathbf{Z} > - < B, dA_\chi(A_\chi^*(B'))\mathbf{Z} >. \qquad (\star)$$

Clearly, the anchor conditions are precisely the conditions (a) and (b) of Proposition 2.2.4. Therefore, it remains to verify condition (c). To this end, let $\rho : G \longrightarrow Aut(L(\mathfrak{g}^*, \mathfrak{g}))$ be the adjoint action and set

$$\rho_\chi : H \times U \longrightarrow Aut(L(\mathfrak{g}^*, \mathfrak{g}))$$
$$(h, p) \mapsto \rho(\chi(h, p)).$$

Then ρ_χ is a groupoid representation. We begin by showing that both sides of condition 2.2.4 (c) are groupoid 1-cocycles for ρ_χ (see Definition 2.1.6).

That the left hand side $P \circ I : H \times U \to L(\mathfrak{g}^*, \mathfrak{g})$ is a groupoid 1-cocycle for ρ_χ is immediate from the fact that $P : U \times G \times U \to L(\mathfrak{g}^*, \mathfrak{g})$ is such a cocycle for ρ.

As for the right hand side, we have to show that the map $\Sigma : H \times U \to L(\mathfrak{g}^*, \mathfrak{g})$ defined by

$$< \alpha, \Sigma(h, p)\beta > = < \lambda_\alpha, Z_\beta > - < \lambda_\beta, Z_\alpha > - < p, [Z_\alpha, Z_\beta] >$$

satisfies

$$\Sigma(hk, p) = \Sigma(h, Ad_{k^{-1}}^* p) + Ad_{\chi(h, Ad_{k^{-1}}^* p)} \Sigma(k, p) Ad_{\chi(h, Ad_{k^{-1}}^* p)}^*.$$

But this follows by a direct calculation which makes successive use of the following three identities

$$< Z_\alpha(hk, p), \lambda > = < Z_\alpha(h, Ad_{k^{-1}}^* p), Ad_{k^{-1}}^* \lambda > + < Z_{Ad_{\chi(h, Ad_{k^{-1}}^* p)}^* \alpha}(k, p), \lambda >$$

$$< \lambda_\alpha(hk, p), Z > = < Ad_k^* \lambda_\alpha(h, Ad_{k^{-1}}^* p), Z >$$

$$< \lambda_\alpha(hk, p), Z > = < \lambda_{Ad_{\chi(h, Ad_{k^{-1}}^* p)}^* \alpha}(k, p), Z > + < p, [Ad_{k^{-1}} Z_\alpha(h, Ad_{k^{-1}}^* p), Z] >$$

We will check the first one and leave the others to the reader. We have

$$< Z_\alpha, \lambda > (hk, p)$$
$$=< \alpha, T_{(hk,p)}(r_{\chi(hk,p)^{-1}} \circ \chi)(0, \lambda) >$$
$$=< \alpha, \frac{d}{dt}\Big|_0 \chi(hk, p + t\lambda)\, \chi(hk, p)^{-1} >$$
$$=< \alpha, \frac{d}{dt}\Big|_0 \left(\chi(h, Ad^*_{k^{-1}}(p + t\lambda))\, \chi(k, p + t\lambda)\, \chi(k, p)^{-1}\, \chi(h, Ad^*_{k^{-1}}p)^{-1}\right) >$$
(by (2.2.2))
$$=< \alpha, \frac{d}{dt}\Big|_0 \left(\chi(h, Ad^*_{k^{-1}}(p + t\lambda))\, \chi(h, Ad^*_{k^{-1}}p)^{-1}\right) >$$
$$+ < \alpha, Ad_{\chi(h, Ad^*_{k^{-1}}p)} \frac{d}{dt}\Big|_0 \left(\chi(h, p + t\lambda)\, \chi(k, p)^{-1}\right) >$$
$$=< Z_\alpha(h, Ad^*_{k^{-1}}p), Ad^*_{k^{-1}}\lambda > + < Z_{Ad^*_{\chi(h, Ad^*_{k^{-1}}p)}\alpha}(k, p), \lambda > .$$

Now, since H is simply connected, by Proposition 7.3 of [**X**] the two groupoid 1-cocycles $P \circ I$ and Σ coincide iff their induced algebroid cocycles are the same. But the latter is equivalent to $(\star)$ above. This concludes the proof. ■

Recall that a Lie algebroid A over a (connected) base B is said to be transitive iff its anchor map $a : A \to TB$ is a surjective submersion. In this case the kernel $Ker\, a$ of a is a Lie algebra bundle [**M1**] called the adjoint bundle of A whose fibers are called the vertex (or isotropy) Lie algebras of A. If A is a transitive Lie algebroid over a contractible base B, it is shown in [**M1**] that A is isomorphic to the trivial Lie algebroid $TB \oplus (B \times \mathfrak{k})$ (Whitney sum), where $\mathfrak{k}$ is the typical fiber of $Ker\, a$; in particular A integrates to a global Lie groupoid isomorphic to $B \times K \times B$ where K is the connected and simply connected Lie group with $Lie(K) = \mathfrak{k}$.

With these facts, we immediately obtain a description of the dual of a dynamical Poisson groupoid over a contractible base.

Let X be a dynamical Poisson groupoid as in Definition 2.2.6 over the contractible base U with embedding of the Hamiltonian unit given by

$$I(h, q) = (Ad^*_{h^{-1}}q, h, q).$$

Let $\iota : \mathfrak{h} \to \mathfrak{g}$ be the inclusion map.

Theorem 3.2.4. (Duality.) *The dual Poisson groupoid X^* of X is isomorphic to the Poisson groupoid $(U \times G' \times U, \{\, ,\, \}_{U \times G' \times U})$ where G' is the connected and simply connected Lie group whose Lie algebra is the vector space $\mathfrak{k} := \{(Z, A) \in \mathfrak{h} \times \mathfrak{g}^* \mid ad^*_Z(q_0) = \iota^* A\}$ for some $q_0 \in U$, equipped with the Lie bracket*

$$[(Z, A), (Z', A')] = \big(-[Z, Z'] - < A, \delta_1 P(\cdot) A' >,$$
$$ad^*_{\iota Z} A' - ad^*_{\iota Z'} A + < A, \partial P(\cdot) A' > \big),$$

and the Poisson bracket $\{\, ,\, \}_{U \times G' \times U}$ is given by Theorem 2.2.5 for a (unique) Poisson groupoid morphism $I' : H \times U \to U \times G' \times U$ and a (unique) groupoid 1-cocycle

$$P' : U \times G' \times U \longrightarrow L(\mathfrak{k}^*, \mathfrak{k})$$

for the adjoint action of $\mathfrak{k}$.

Proof. For the first part, observe that the anchor map of $A(X)^*$

$$a_*(q, Z, A) = (q, -ad_Z^* q + \iota^* A)$$

is a surjective submersion since ι is injective. Thus $A(X)^*$ is transitive and therefore, by Mackenzie's theorem, it is isomorphic to the trivial Lie algebroid $A' = U \times \mathfrak{h}^* \times (Ker\, a_*)_{q_0}$. Now the fiber $(Ker\, a_*)_{q_0}$ is the vector space $\mathfrak{k}$ equipped with the Lie bracket given by the restriction of the bracket of sections of $A(X)^*$ of Proposition 3.2.1 (with $K(q)Z = ad_Z^* q$ and $A(q)Z = \iota Z$). Hence the claim.

For the second part, let

$$\tau : A(X)^* \longrightarrow A'$$

be the (base preserving) trivializing isomorphism of Mackenzie's theorem, and denote by

$$T : X^* \longrightarrow U \times G' \times U$$

the unique groupoid isomorphism such that $A(T) = \tau$. We may thus transport the Poisson groupoid structure of X^* to $U \times G' \times U$ by setting

$$\{f, g\}_{U \times G' \times U} = \{f \circ T, g \circ T\}_{X^*} \circ T^{-1}.$$

We now show that there is a (base preserving) Poisson groupoid morphism

$$I' : H \times U \longrightarrow U \times G' \times U,$$

where $H \times U$ is the Hamiltonian unit.

Consider the Poisson groupoid morphism (this is the anchor map of X)

$$J : U \times G \times U \longrightarrow U \times U : (p, x, q) \mapsto (p, q)$$

where $U \times U$ is the coarse groupoid of Example 3.1.2. Its induced Lie bialgebroid morphism

$$A(J) : U \times \mathfrak{g} \times \mathfrak{h}^* \longrightarrow U \times \mathfrak{h}^*$$

is of course just the anchor map a of $A(X)$. By the lifting property of Lie algebroid morphisms, and Proposition 3.2.3 above, the dual morphism

$$A(J)^* = a^* : U \times \mathfrak{h} \longrightarrow U \times \mathfrak{g}^* \times \mathfrak{h}$$

may be lifted uniquely to a (base preserving) Poisson groupoid morphism

$$J^* : H \times U \longrightarrow X^*.$$

Thus $I' = T \circ J^*$ is the sought-for Poisson groupoid morphism. The uniqueness of I' and P' now follows from the uniqueness of the Poisson structure of a (suitably simply connected) Poisson groupoid with prescribed tangent Lie bialgebroid [**MX2**]. Hence the claim. ∎

Caveat

We will see in section 5 that, for $\mathfrak{h} \neq 0$, even when X is coboundary with **constant** r-matrix, the vertex group G' is different from the Poisson Lie group dual to G equipped with the Sklyanin bracket $\{\,,\,\}_{(R,-R)}$.

We close this subsection with a description of natural Poisson quotients associated with Theorem 3.2.4. We now assume that the contractible set U contains 0.

Let X be as in Theorem 2.2.5 with the map $H \longrightarrow G : h \mapsto \chi(h, 0)$ one to one. Consider the restriction of the left Hamiltonian action

$$\phi^- : H \times X \longrightarrow X : (h, (p, x, q)) \mapsto (Ad^*_{h^{-1}} p, \chi(h, p)x, q)$$

to $\alpha^{-1}(0) = \{0\} \times G \times U$:

$$\phi^- : H \times G \times U \longrightarrow G \times U : (h, x, q) \mapsto (\chi(h, 0)x, q).$$

Let $\pi : G \times U \to G/H \times U : (x, p) \mapsto (\overline{x}, p)$ be the canonical projection.

Proposition 3.2.5. (Hamiltonian reduction.) *The Poisson bracket $\{\,,\,\}_{red.}$ of the reduced space $\alpha^{-1}(0)/H \simeq G/H \times U$ vanishes at $(\overline{1}, 0)$. Its linearization at $(\overline{1}, 0)$ coincides with the vertex Lie algebra of $A(X)^*$ at 0.*

Proof. We have to calculate the Poisson bracket $\{f, g\}(0, x, q)$ of two functions $f, g \in C^\infty(X)$ whose restriction to $\{0\} \times G \times U$ is H-invariant.

Since H is connected, the restriction of f to $\{0\} \times G \times U$ is H-invariant if and only if

$$\frac{d}{dt}\bigg|_{t=0} f(0, \chi(e^{tZ}, 0)x, q) = 0, \quad \forall Z \in \mathfrak{h},$$

that is, if and only if

$$A^*_\chi(0) Df(0, x, q) = 0 \text{ for all } x \in G, q \in U.$$

Thus (see Theorem 2.2.5)

$$\{f, g\}(0, x, q) = - <q, [\delta_2 f, \delta_2 g]> - <A_\chi(q)\delta_2 f, D'g>$$
$$+ <A_\chi(q)\delta_2 g, D'f> + <Df, P(0, x, q)Dg>.$$

Now, $P(0, 1, 0) = 0$ and $A^*_\chi(0)D'f(0, 1, 0) = A^*_\chi(0)Df(0, 1, 0) = 0$. Therefore,

$$\{f, g\}(0, 1, 0) = 0.$$

Set $Z = \delta_2 f, Z' = \delta_2 g, A = Df, A' = Dg$, all evaluated at $(0, 1, 0)$. A direct calculation then gives

$$d\{f, g\}(0, 1, 0)(0, Y, \lambda) = - <\lambda, [Z, Z']> - <dA_\chi(0)(\lambda) Z, A'>$$
$$+ <dA_\chi(0)(\lambda)Z', A> + <ad^*_{A_\chi(0)Z} A', Y>$$
$$- <ad^*_{A_\chi(0)Z'} A, Y> + <A, (\partial P(X) + \delta_2 P(\lambda))A'>.$$

To conclude, observe that this coincides with the restriction of the Lie bracket of Proposition 3.2.1 (with $K(q)Z = ad^*_Z q$ and $A = A_\chi$) to the kernel $Ker\, a_*$ at 0. ∎

Proposition 3.2.5 provides in some sense an indirect Poisson integration of the vertex Lie algebra $(Ker\, a_*)_0$ of $A(X)^*$ by the natural quotient space $G/H \times U$. For X dynamical, combining Proposition 3.2.5 with Theorem 3.2.4 then gives a reduced vertex diagram reminiscent of the Poisson Lie group duality of Drinfel'd.

Let $X = U \times G \times U$ be a dynamical Poisson groupoid as in Definition 2.2.6 with dual Poisson groupoid $X^* \simeq U \times G' \times U$. Assume that the map $H \to G' : h \mapsto$

$\chi'(h, 0)$ is one to one. Denote the units of G and G' by 1 and equip both spaces $G/H \times U$, $G'/H \times U$ with the Poisson brackets obtained via Poisson reduction.

Let $\mathfrak{h}^\perp \subset \mathfrak{g}^*$ be the annihilator of $\mathfrak{h}$.

Proposition 3.2.6. (Reduced duality diagram) *We have the diagram*

$$
\begin{array}{ccc}
X^* & & X \\
{\scriptstyle red.}\searrow & & \swarrow{\scriptstyle red.} \\
G'/H \times U & & G/H \times U \\
& \searrow \swarrow & \\
& linearize\ at\ (\overline{1}, 0) & \\
& \swarrow \searrow & \\
A(X)^* \supset (Ker\ a_*)_0 = \mathfrak{h} \times \mathfrak{h}^\perp & & \mathfrak{g} = (Ker\ a)_0 \subset A(X)
\end{array}
$$

In case H is reduced to its unit, this diagram reduces to that of Drinfel'd's duality for Poisson Lie groups. ∎

CHAPTER 4

An Explicit Case Study of Duality

In [**EV**], Etingof and Varchenko obtained, among other things, a classification of solutions of the CDYBE for pairs $(\mathfrak{g}, \mathfrak{h})$ of Lie algebras, where $\mathfrak{g}$ is simple, and $\mathfrak{h} \subset \mathfrak{g}$ is a Cartan subalgebra. Our purpose in this section is to give an explicit study of duality for the corresponding class of coboundary dynamical Poisson groupoids.

We begin by recalling the general form of these dynamical r-matrices.

First, let us fix some notation. Let $\mathfrak{g}$ be a complex simple Lie algebra with Killing form $(\,,\,)$, $\mathfrak{h} \subset \mathfrak{g}$ a Cartan subalgebra, and $\mathfrak{g} = \mathfrak{h} \oplus \sum_{\alpha \in \Delta} \mathfrak{g}_\alpha$ the root space decomposition. We let Δ^s be a fixed simple system of roots and denote by $\Delta^\pm$ the corresponding positive/negative system. For any positive root $\alpha \in \Delta^+$, we choose root vectors $e_\alpha \in \mathfrak{g}_\alpha$ and $e_{-\alpha} \in \mathfrak{g}_{-\alpha}$ which are dual with respect to $(\,,\,)$ so that $[e_\alpha, e_{-\alpha}] = h_\alpha$. We also fix an orthonormal basis $(x_i)_{1 \leq i \leq rank(\mathfrak{g})}$ of $\mathfrak{h}$. Lastly, for a subset of simple roots $\Gamma \subset \Delta^s$, we will denote the root span of Γ by $<\Gamma> \subset \Delta$ and set $\overline{\Gamma}^\pm = \Delta^\pm \backslash <\Gamma>^\pm$.

For any subset $\Gamma \subset \Delta^s$, we give the $ad_{\mathfrak{h}}$-equivariant solutions of the mDYBE (see (2.1.1), (2.1.2)) associated with the triple $(\mathfrak{g}, \mathfrak{h}, \Gamma)$ as (cf.[**EV**]):

$$R(q)B = \sum_{i,j} C_{ij}(q) <x_j, B> x_i + \sum_{\alpha \in \Delta} \phi_\alpha(q) <e_{-\alpha}, B> e_\alpha \qquad (4.1)$$

where

$$\phi_\alpha(q) = \frac{1}{2} \text{ for } \alpha \in \overline{\Gamma}^+, \quad \phi_\alpha(q) = -\frac{1}{2} \text{ for } \alpha \in \overline{\Gamma}^-$$

$$\phi_\alpha(q) = \frac{1}{2} coth(\frac{(\alpha, q - \mu)}{2}) \text{ for } \alpha \in <\Gamma>,$$

and where $\sum_{i,j} C_{ij} dq^i \otimes dq^j$ is any closed meromorphic 2-form on $\mathfrak{h}^*$ and $\mu \in \mathfrak{h}^*$ is arbitrary.

We will denote by U the domain of analyticity of R and let G be the connected and simply-connected Lie group with $Lie(G) = \mathfrak{g}$. Note that U is trivially Ad^*_H-invariant as H is abelian, hence we can consider the coboundary dynamical Poisson groupoid $X = U \times G \times U$ associated to R. Our immediate goal is to construct an explicit trivialization of the dynamical Lie algebroid $A(X)^* \simeq U \times \mathfrak{h}^* \times \mathfrak{g}^*$. Note that, as U is not contractible, this is not guaranteed by Mackenzie's theorem. In what follows, we will make the identification $\mathfrak{g}^* \simeq \mathfrak{g}$ using the Killing form $(\,,\,)$. Then we have $\mathfrak{h}^\perp \simeq \mathfrak{n} := \sum_{\alpha \in \Delta} \mathfrak{g}_\alpha$, and the Lie bracket between the sections of the

26

dynamical Lie algebroid takes the form

$$[(Z, B),(Z', B')]_*(q)$$
$$= (dZ'(q)\Pi_\mathfrak{h} B(q) - dZ(q)\Pi_\mathfrak{h} B'(q) - (dR(q)(.)B(q), B'(q)),$$
$$- dB(q)\Pi_\mathfrak{h} B'(q) + dB'(q)\Pi_\mathfrak{h} B(q)$$
$$- [R(q)(B(q)) + Z(q), B'(q)] - [B(q), R(q)(B'(q)) + Z'(q)]), \quad (4.2)$$

where $\Pi_\mathfrak{h}$ is the projection map onto $\mathfrak{h}$ relative to the direct sum decomposition $\mathfrak{g} = \mathfrak{h} \oplus \mathfrak{n}$.

We will begin our construction with a description of the vertex Lie algebra $\mathcal{V}_q = (Ker\, a_*)_q = \mathfrak{h} \times \mathfrak{h}^\perp \simeq \mathfrak{h} \times \mathfrak{n}$ of A^* at $q \in U$. To do so, let us introduce the following Lie subalgebras of $\mathfrak{g}$ associated with $\Gamma \subset \Delta^s$:

$$\mathfrak{h}_\Gamma := <(h_\gamma)_{\gamma \in \Gamma}>_{\mathbf{C}},$$
$$\mathfrak{h}_\Gamma^\perp := \text{ the orthogonal complement of } \mathfrak{h}_\Gamma \text{ in } \mathfrak{h} \text{ w.r.t. } (\,,\,)_{|\mathfrak{h} \times \mathfrak{h}},$$
$$\mathfrak{l}_\Gamma := \mathfrak{h}_\Gamma \oplus <(e_\alpha)_{\alpha \in <\Gamma>}>_{\mathbf{C}} \text{ the Levi factor },$$
$$\overline{\mathfrak{n}}_\Gamma^\pm := <(e_\alpha)_{\alpha \in \overline{\Gamma}^\pm}>_{\mathbf{C}} \text{ the nilpotent radicals }.$$

Clearly, $[\mathfrak{h}_\Gamma, \overline{\mathfrak{n}}_\Gamma^\pm] \subset \overline{\mathfrak{n}}_\Gamma^\pm$ and $[\mathfrak{l}_\Gamma, \overline{\mathfrak{n}}_\Gamma^\pm] \subset \overline{\mathfrak{n}}_\Gamma^\pm$.

If $\mathfrak{g}_1, \mathfrak{g}_2$ are two Lie algebras, we denote by $\mathfrak{g}_1 \ominus \mathfrak{g}_2$ the vector space $\mathfrak{g}_1 \oplus \mathfrak{g}_2$ equipped with the Lie bracket $[x_1 + x_2, y_1 + y_2] = [x_1, y_1] - [x_2, y_2]$. Let $\mathfrak{J}_\Gamma = \mathfrak{h}_\Gamma^\perp \ltimes (\overline{\mathfrak{n}}_\Gamma^+ \ominus \overline{\mathfrak{n}}_\Gamma^-)$ be the semidirect product Lie algebra where $\mathfrak{h}_\Gamma^\perp$ acts on each summand of the anti-direct sum by the adjoint action of $\mathfrak{g}$. Set

$$\mathfrak{g}' := \mathfrak{l}_\Gamma \ltimes \mathfrak{J}_\Gamma$$

where $\mathfrak{l}_\Gamma$ also acts on $\mathfrak{J}_\Gamma$ by the adjoint action of $\mathfrak{g}$.

Proposition 4.1. *Let $q \in U$. Then the map $\psi(q) : \mathcal{V}_q \longrightarrow \mathfrak{g}'$ defined by*

$$\psi(q)(0, e_\alpha) = \frac{-1}{2 sinh\left(\frac{(\alpha, q-\mu)}{2}\right)} e_\alpha \quad \text{for } \alpha \in <\Gamma>,$$
$$\psi(q)(0, e_\alpha) = -e^{\mp \frac{1}{2}(\alpha, q-\mu)} e_\alpha \quad \text{for } \alpha \in \overline{\Gamma}^\pm,$$
$$\psi(q)(Z, 0) = -Z \quad \text{for all } Z \in \mathfrak{h},$$

is an isomorphism of Lie algebras.

Proof. The Lie bracket of $\mathcal{V}_q = \mathfrak{h} \times \mathfrak{n}$ can be calculated from (4.2) and we have

$$[(Z, n), (Z', n')]_* = \big(-(dR(q)(.)n, n'), -[R(q)n + Z, n']$$
$$- [n, R(q)n' + Z'] \big).$$

Writing $n = \sum_{\alpha \in \Delta} n^\alpha e_\alpha$ and similarly for n', we have

$$(dR(q)(\Lambda)n, n') = \sum_{\alpha \in <\Gamma>} d\phi_\alpha(q)(\Lambda) n^\alpha n'^{-\alpha}$$
$$= \sum_{\alpha \in <\Gamma>} (\frac{1}{4} - \phi_\alpha(q)^2)(\alpha, \Lambda) n^\alpha n'^{-\alpha}.$$

A direct calculation then gives the following Lie bracket relations:

$[(0, e_\alpha), (0, e_\beta)]_* = (0, -(\phi_\alpha(q) + \phi_\beta(q))[e_\alpha, e_\beta])$ for $\alpha \in <\Gamma>, \beta \in \Delta, \alpha + \beta \neq 0$,

$[(0, e_\alpha), (0, e_{-\alpha})]_* = (-(\frac{1}{4} - \phi_\alpha(q)^2)[e_\alpha, e_{-\alpha}], 0)$ for $\alpha \in <\Gamma>$,

$[(0, e_\alpha), (0, e_\beta)]_* = (0, 0)$ for $\alpha \in \overline{\Gamma}^+, \beta \in \overline{\Gamma}^-$,

$[(0, e_\alpha), (0, e_\beta)]_* = (0, -[e_\alpha, e_\beta])$ for $\alpha, \beta \in \overline{\Gamma}^+$

$[(0, e_\alpha), (0, e_\beta)]_* = (0, +[e_\alpha, e_\beta])$ for $\alpha, \beta \in \overline{\Gamma}^-$,

$[(Z, 0), (0, e_\alpha)]_* = (0, -\alpha(Z)e_\alpha)$ for all $\alpha \in \Delta, Z \in \mathfrak{h}$

$[(Z, 0), (Z', 0)]_* = (0, 0), \quad Z, Z' \in \mathfrak{h}$.

where the bracket $[,]$ on the right hand side is that of $\mathfrak{g}$. After rescaling the basis of $\mathfrak{n}$ by setting

$$E_\alpha(q) = 2\sinh(\frac{(\alpha, q-\mu)}{2})e_\alpha, \alpha \in <\Gamma>; \quad E_\alpha(q) = e^{\pm\frac{1}{2}(\alpha, q-\mu)}e_\alpha, \alpha \in \overline{\Gamma}^\pm,$$

the above relations yield

$[(0, E_\alpha(q)), (0, E_\beta(q))]_* = (0, -N_{\alpha,\beta}E_{\alpha+\beta}(q))$ for $\alpha \in <\Gamma>, \beta \in \Delta, \alpha + \beta \neq 0$,

$[(0, E_\alpha(q)), (0, E_{-\alpha}(q))]_* = (-[e_\alpha, e_{-\alpha}], 0)$ for $\alpha \in <\Gamma>$,

$[(0, E_\alpha(q)), (0, E_\beta(q))]_* = (0, 0)$ for $\alpha \in \overline{\Gamma}^+, \beta \in \overline{\Gamma}^-$,

$[(0, E_\alpha(q)), (0, E_\beta(q))]_* = (0, -N_{\alpha,\beta}E_{\alpha+\beta}(q))$ for $\alpha, \beta \in \overline{\Gamma}^+$

$[(0, E_\alpha(q)), (0, E_\beta(q))]_* = (0, +N_{\alpha,\beta}E_{\alpha+\beta}(q))$ for $\alpha, \beta \in \overline{\Gamma}^-$,

$[(Z, 0), (0, E_\alpha(q))]_* = (0, -\alpha(Z)E_\alpha(q))$ for all $\alpha \in \Delta, Z \in \mathfrak{h}$,

$[(Z, 0), (Z', 0)]_* = (0, 0) \quad Z, Z' \in \mathfrak{h}$,

where $N_{\alpha,\beta}$ are the structure constants of $\mathfrak{g}$.

We will check the first Lie bracket above with $\beta \in \overline{\Gamma}^-$; the others are similar. For a root $\gamma \in \Delta$, set $x_\gamma = (\gamma, q-\mu)$. We have $\phi_\alpha(q) + \phi_\beta(q) = \frac{1}{2}\coth(\frac{x_\alpha}{2}) - \frac{1}{2} = \frac{1}{e^{x_\alpha}-1}$, $E_\alpha(q) = e^{\frac{x_\alpha}{2}}(1-e^{-x_\alpha})e_\alpha$, and $E_\beta(q) = e^{\frac{-x_\beta}{2}}e_\beta$. Thus,

$$[(0, E_\alpha(q)), (0, E_\beta(q))]_* = (0, -\frac{e^{\frac{x_\alpha}{2}}(1-e^{-x_\alpha})e^{\frac{-x_\beta}{2}}}{(e^{x_\alpha}-1)}N_{\alpha,\beta}e_{\alpha+\beta}).$$

Now, if $\alpha + \beta$ is a root, then it belongs to $\overline{\Gamma}^-$; thus $e_{\alpha+\beta} = e^{\frac{x_\alpha+x_\beta}{2}}E_{\alpha+\beta}(q)$, and this immediately gives the assertion.

The Lie bracket relations above show that the structure constants of $\mathcal{V}_q$ in the basis $((x_i, 0) 1 \leq i \leq rank(\mathfrak{g}); (0, E_\alpha(q)), \alpha \in \Delta)$ are opposite to those of $\mathfrak{g}'$. Therefore the map $\psi : \mathcal{V}_q \longrightarrow \mathfrak{g}'$ defined by $(Z, 0) \mapsto -Z$ and $(0, E_\alpha(q)) \mapsto -e_\alpha$ is an isomorphism of Lie algebras. Hence the claim. ∎

Corollary 4.2. *The map $\widetilde{\psi} : U \times \mathfrak{g}' \longrightarrow Ker\, a_* : (q, \xi) \longrightarrow (q, -\Pi_\mathfrak{h}\xi, \psi(q)^{-1}(\Pi_\mathfrak{n}\xi))$ is an isomorphism between the trivial Lie algebra bundle $U \times \mathfrak{g}'$ and the adjoint bundle $Ker\, a_*$. Here, $\Pi_\mathfrak{h}$ and $\Pi_\mathfrak{n}$ are the projections relative to the direct sum decomposition $\mathfrak{g} = \mathfrak{h} \oplus \mathfrak{n}$.* ∎

Let us briefly comment on the vertex isomorphism of Proposition 4.1. Let $\mathfrak{n}^\pm = <(e_\alpha)_{\alpha \in \Delta^\pm}>_{\mathbf{C}}$. If $\Gamma \subset \Delta^s$ is the empty set, the vertex Lie algebra $\mathcal{V}_q$, $q \in U$, of $A(X)^*$ is isomorphic to $\mathfrak{h} \ltimes (\mathfrak{n}^+ \ominus \mathfrak{n}^-)$, which is reminiscent of (although not identical to) the Lie algebra dual of $\mathfrak{g}$ equipped with the standard constant r-matrix (see also Example 5.1.8). If $\Gamma = \Delta^s$, we have $\mathcal{V}_q \simeq \mathfrak{g}' = \mathfrak{l}_\Delta = \mathfrak{g}$. For a general subset $\Gamma \subset \Delta^s$ the vertex Lie algebra $\mathcal{V}_q \simeq \mathfrak{g}'$ is seen to naturally intertwine the Levi factor $\mathfrak{l}_\Gamma$ with the summand $\mathfrak{J}_\Gamma$ which is again reminiscent of the Lie algebra dual of $\mathfrak{g}$ equipped with the standard constant r-matrix.

Our next step is to construct a flat connection $\theta_* : TU \simeq U \times \mathfrak{h}^* \longrightarrow U \times \mathfrak{h} \times \mathfrak{g}^*$ satifying the condition $[\theta_*(\lambda), \widetilde{\psi}(\xi)]_* = \widetilde{\psi}(d\xi \cdot \lambda)$ for $\lambda : U \longrightarrow \mathfrak{h}^*$ and $\xi : U \longrightarrow \mathfrak{g}'$. To simplify notation, we will identify the elements $(Z, n) \in \mathfrak{h} \times \mathfrak{n} \simeq \mathcal{V}_q$ of the vertex Lie algebra with $Z + n \in \mathfrak{g}$ from now onwards.

Let $C^\# : U \to L(\mathfrak{h}^*, \mathfrak{h})$ be the map defined by $C^\#(q)\lambda = \sum_{i,j} C_{ij}(q)\lambda(x_j)x_i$. We will seek θ_* in the form $\theta_*(q, \lambda) = (q, f(q)\lambda, \lambda)$, where $f : U \longrightarrow L(\mathfrak{h}^*, \mathfrak{h})$. By definition, θ_* is a flat connection if and only if $\theta_*[\lambda, \lambda'] = [\theta_*(\lambda), \theta_*(\lambda')]_*$ for $\lambda, \lambda' : U \longrightarrow \mathfrak{h}^*$. By using Eqn. (4.2), a straightforward calculation shows that this is equivalent to the following two conditions:

(1) $df(q)(\lambda'(q))\lambda(q) - df(q)(\lambda(q))\lambda'(q) = -(dR(q)(\,.\,)\lambda(q), \lambda'(q))$

(2) $[R(q)(\lambda(q)), \lambda'(q)] + [\lambda(q), R(q)(\lambda'(q))] = 0$

for $q \in U$.

On the other hand, the condition $[\theta_*(\lambda), \widetilde{\psi}(\xi)]_* = \widetilde{\psi}(d\xi \cdot \lambda)$ is equivalent to

(3) $(dR(q)(\,.\,)\lambda(q), n) = 0$

(4) $d\psi^{-1}(q)\lambda(q)(n) - [f(\lambda)(q), \psi^{-1}(q)(n)]$
$\quad - \big([R(q)(\lambda(q)), \psi^{-1}(q)(n)] + [\lambda(q), R(q)(\psi^{-1}(q)(n))]\big) = 0,$

for $n \in \mathfrak{n}$ and $q \in U$.

From the properties of R, (2) and (3) are immediately seen to hold. We now examine condition (4). Set $\psi(q)(e_\alpha) = \psi_\alpha(q)e_\alpha$, for all $\alpha \in \Delta$, then $d\psi^{-1}(q)\lambda(q)(e_\alpha) = \phi_\alpha(q)\psi_\alpha(q)^{-1}(\lambda(q), \alpha)e_\alpha$. Meanwhile, it is easy to check that

$$[f(\lambda)(q), \psi^{-1}(q)e_\alpha] = \psi_\alpha(q)^{-1}\alpha(f(\lambda(q)))e_\alpha$$
$$\big([R(q)(\lambda(q)), \psi^{-1}(q)e_\alpha] + [\lambda(q), R(q)(\psi^{-1}(q)e_\alpha)]\big)$$
$$= \big(\psi_\alpha(q)^{-1}(\alpha(C^\#(q)(\lambda(q))) + (\alpha, \lambda(q))\phi_\alpha(q)\big)e_\alpha.$$

Therefore, condition (4) is equivalent to

$$\alpha\big(f(\lambda)(q) + C^\#(q)(\lambda(q))\big) = 0, \text{ for all } \alpha \in \Delta,$$

that is to $f = -C^\#$. Finally, inserting $f = -C^\#$ into condition (1) shows that it is trivially satisfied as it is equivalent to the closedness of the 2-form $\sum_{i,j} C_{ij} dq^i \otimes dq^j$. Hence we have

Proposition 4.3. *The map*

$$\theta_* : TU \simeq U \times \mathfrak{h}^* \longrightarrow U \times \mathfrak{h} \times \mathfrak{g}^*$$
$$(q, \lambda) \mapsto (q, -C^\#(q)\lambda, \lambda)$$

is a flat connection on $U \times \mathfrak{h} \times \mathfrak{g}^*$ satisfying $[\theta_*(\lambda), \widetilde{\psi}(\xi)]_* = \widetilde{\psi}(d\xi \cdot \lambda)$ for $\lambda : U \longrightarrow \mathfrak{h}^*$ and $\xi : U \longrightarrow \mathfrak{g}'$. ∎

Theorem 4.4. (Trivialization) *Let $A' := U \times \mathfrak{h}^* \times \mathfrak{g}'$ be the trivial Lie algebroid over U (see Proposition 3.2.1). Then the (bijective) bundle map*

$$\sigma : A' \longrightarrow U \times \mathfrak{h} \times \mathfrak{g}^*$$
$$(q, \lambda, \xi) \mapsto \theta_*(q, \lambda) + \widetilde{\psi}(q, \xi)$$

is an isomorphism of Lie algebroids with inverse given by $\tau(q, Z, \lambda + n) = (q, \lambda, -C^\#(q)\lambda - Z + \psi(q)n)$. In particular, the coboundary dynamical Lie algebroid $A(X)^* \simeq U \times \mathfrak{h} \times \mathfrak{g}^*$ is integrable.

Proof. This is clear from the properties of θ_* and the fact that $\widetilde{\psi}$ is an isomorphism of Lie algebra bundles. ∎

Remark 4.5. In general, it can be shown that any coboundary dynamical Lie algebroid $A(X)^*$ associated to a solution of (mDYBE) is integrable via its embedding into the Lie algebroid direct sum $A(X) \underset{TU}{\oplus} A(X)$ [**L1**]. Indeed, this can be used as the starting point of an alternative proof of Theorem 4.4 by using the results in [**L2**]. However, our proof here is direct and does not depend on the development of excessive machinery.

In what follows, we let U' be a connected and simply-connected open subset of U and we consider the coboundary Poisson groupoid $X(U') = U' \times G \times U'$ associated to R. We also let G' be the connected and simply connected Lie group with $Lie(G') = \mathfrak{g}'$ and denote by

$$T : X(U')^* \longrightarrow X' = U' \times G' \times U'$$

the unique (base preserving) Lie groupoid isomorphism such that $A(T) = \tau|_{U' \times \mathfrak{h} \times \mathfrak{g}^*}$.

If we define the Poisson bracket on X' by

$$\{f, g\}_{X'} = \{f \circ T, g \circ T\}_{X(U')^*} \circ T^{-1},$$

then $(X', \{\,,\,\}_{X'})$ and $(X(U'), \{\,,\,\}_{X(U')})$ are Poisson groupoids in duality (see Definition 3.1.1) and T is an isomorphism of Poisson groupoids.

The following theorem characterizes the Poisson groupoid $(X', \{\,,\,\}_{X'})$.

Let $j : \mathfrak{h} \longrightarrow \mathfrak{g}' : Z \mapsto Z$ be the inclusion.

Theorem 4.6. *The Poisson groupoid $(X', \{\,,\,\}_{X'})$ is of dynamical type with Poisson bracket*

$$\{f, g\}_{X'}(p, u, q) = <p, [\delta_1 f, \delta_1 g]> - <q, [\delta_2 f, \delta_2 g]>$$
$$- <j\delta_1 f, Dg> - <j\delta_2 f, D'g>$$
$$+ <j\delta_1 g, Df> + <j\delta_2 g, D'f>$$
$$+ <Df, P'(p, u, q)Dg>,$$

4. AN EXPLICIT CASE STUDY OF DUALITY

where $P' : U' \times G' \times U' \longrightarrow L(\mathfrak{g}'^*, \mathfrak{g}')$ is the unique skew symmetric groupoid cocycle whose tangent cocycle $P'_*(q, \lambda, Z + n) := -\delta_1 P'(\lambda) + \partial P'(Z + n)$ is given by

$$(n_1, \delta_1 P'(\lambda) n_2) = (\Pi_{\mathfrak{h}}[\psi^* n_1, \psi^* n_2], \lambda)$$
$$(\lambda_1, \delta_1 P'(\lambda) n_2) = 0$$
$$(\lambda_1, \delta_1 P'(\lambda) \lambda_2) = (dC^{\#}(\lambda_2) \lambda_1 - dC^{\#}(\lambda_1) \lambda_2, \lambda)$$
$$(n_1, \partial P'(n) n_2) = -(\Pi_{\mathfrak{n}}[\psi^* n_1, \psi^* n_2], \psi^{-1} n)$$
$$(\lambda_1, \partial P'(n) n_2) = -(d(\psi^{*-1})(\lambda_1) \psi^* n_2, n) - ([C^{\#} \lambda_1, n_2], n)$$
$$(\lambda_1, \partial P'(n) \lambda_2) = 0$$
$$\partial P'(Z) = 0.$$

Here, $\Pi_{\mathfrak{h}}$ and $\Pi_{\mathfrak{n}}$ are the projections relative to the direct sum decomposition $\mathfrak{g} = \mathfrak{h} \oplus \mathfrak{n}$, $n_1, n_2, n \in \mathfrak{n}$, $Z \in \mathfrak{h}$, and $\lambda_1, \lambda_2, \lambda \in \mathfrak{h}^* \simeq \mathfrak{h}$, and the differentials of P' are taken at $(q, 1, q)$.

Proof. For the sake of clarity, we will begin by repeating the argument of Theorem 3.2.4 here. Consider the Poisson groupoid morphism

$$J : X \longrightarrow U' \times U' : (p, h, q) \mapsto (p, q),$$

with induced Lie bialgebroid morphism $A(J)$. Applying Proposition 3.2.3 to the dual morphism

$$A(J)^* : U' \times \mathfrak{h} \longrightarrow U' \times \mathfrak{h} \times \mathfrak{g}^*$$

we infer the existence of a (base preserving) Poisson groupoid morphism

$$J^* : H \times U' \longrightarrow X^*$$

and so of a morphism

$$I' = T \circ J^* : H \times U' \longrightarrow U' \times G' \times U'.$$

Now, (see (2.2.1)) I' is necessarily of the form

$$(h, p) \mapsto (p, \chi'(h, p), p),$$

for some groupoid morphism $\chi' : H \times U' \longrightarrow G'$ with tangent map $A(\chi') : U' \times \mathfrak{h} \longrightarrow \mathfrak{g}'$. Therefore, the Poisson bracket $\{\, ,\, \}_{X'}$ is given by Theorem 2.2.5 for $A_{\chi'}$ and some groupoid 1-cocycle $P' : X' \to L(\mathfrak{g}'^*, \mathfrak{g}')$.

Denote the Lie algebroid of $U' \times G' \times U'$ by A' and let $(A', [\,,\,]', a'; A'^*, [\,,\,]'_*, a'_*)$ be the Lie bialgebroid structure of Proposition 3.2.1. For A'^*, we have $K = 0$ since $\mathfrak{h}$ is abelian, and $A = A_{\chi'}$. The Poisson bracket $\{\, ,\, \}_{X'}$ is now uniquely determined by the duality requirement (see Definition 3.1.1) that the trivialization map τ of Theorem 4.4 be a Lie bialgebroid isomorphism from $(A(X)^*, A(X)^-)$ to (A', A'^*), that is, by the condition that the map

$$\tau^* : A'^* = U' \times \mathfrak{h} \times \mathfrak{g}'^* \longrightarrow A(X)^- = U' \times \mathfrak{h}^* \times \mathfrak{g}$$
$$(q, Z, \lambda + n) \mapsto (q, -\lambda, C^{\#}(\lambda) + Z + \psi^*(n))$$

satisfies

$$-a_{A(X)} \tau^* = a'_*,$$
$$\tau^*[(Z_1, \lambda_1 + n_1), (Z_2, \lambda_2 + n_2)]'_* = -[\tau^*(Z_1, \lambda_1 + n_1), \tau^*(Z_2, \lambda_2 + n_2)]_{A(X)}.$$

Now (see Proposition 3.2.1) the anchor condition is equivalent to $A_{\chi'}(q)(Z) = j(Z) = Z$ so the groupoid morphism $\chi' : H \times U' \longrightarrow G'$ is just the inclusion of H into G'. Therefore (see Definition 2.2.6) X' is of dynamical type.

It remains to show that the morphism property is equivalent to the stated defining equations for P'_*. We will illustrate the identities for $\delta_1 P'$ and $\partial P'$ above by calculating

$$d((\lambda_1, Y_1), (\lambda_2, Y_2))$$
$$= \tau^*[\tau^{*-1}(\lambda_1, Y_1), \tau^{*-1}(\lambda_2, Y_2)]'_* + [(\lambda_1, Y_1), (\lambda_2, Y_2)]_{A(X)},$$

in two cases, leaving the others to the reader.

By Proposition 3.2.1,

$$< d((0, n_1), (0, n_2), (Z, \lambda + n) >$$
$$=< [(0, \psi^{*-1}(n_1)), (0, \psi^{*-1}(n_2))]'_*, (\lambda, -C^\#(\lambda) - Z + \psi(n)) >$$
$$+ < [(0, n_1), (0, n_2)]_{A(X)}, (Z, \lambda + n) >$$
$$= -\big(\psi^{*-1}(n_1), \delta_1 P'(\lambda)\psi^{*-1}(n_2)\big)$$
$$+ \big(\psi^{*-1}(n_1), \partial P'(-C^\#(\lambda) - Z + \psi(n))\psi^{*-1}(n_2)\big) + \big([n_1, n_2], \lambda + n\big).$$

Thus $d((0, n_1), (0, n_2)) = 0$ if and only if $(n_1, \delta_1 P'(\lambda)n_2)$ and $(n_1, \partial P'(n)n_2)$ are as stated and $(n_1, \partial P'(Z)n_2) = 0$.

Next,

$$< d((\lambda_1, 0), (\lambda_2, 0)), (Z, \lambda + n) >$$
$$=< [C^\#(\lambda_1), -\lambda_1), (C^\#(\lambda_2), -\lambda_2)]'_*, (\lambda, -C^\#(\lambda) - Z + \psi(n)) >$$
$$+ < [(\lambda_1, 0), (\lambda_2, 0)]_{A(X)}, (Z, \lambda + n) >$$
$$= \big(- d(C^\#(\lambda_2))\lambda_1 + d(C^\#(\lambda_1))\lambda_2, \lambda\big) - \big(\lambda_1, \delta_1 P'(\lambda)\lambda_2\big)$$
$$- \big(d\lambda_2\lambda_1 - d\lambda_1\lambda_2, C^\#(\lambda) + Z\big) - \big(\lambda_2, [C^\#(\lambda_1), \psi(n)]\big)$$
$$+ \big(\lambda_1, [C^\#(\lambda_2), \psi(n)]\big) + \big(\lambda_1, \partial P'(-C^\#(\lambda) - Z + \psi(n))\lambda_2\big)$$
$$+ < [(\lambda_1, 0), (\lambda_2, 0)]_{A(X)}, (Z, \lambda + n) >$$
$$= \big(- dC^\#(\lambda_1)\lambda_2 + dC^\#(\lambda_2)\lambda_1, \lambda\big) - \big(\lambda_1, \delta_1 P'(\lambda)\lambda_2\big)$$
$$- \big(\lambda_2, [C^\#\lambda_1, \psi(n)]\big) + \big(\lambda_1, [C^\#\lambda_2, \psi(n)]\big)$$
$$+ \big(\lambda_1, \partial P'(-C^\#\lambda - Z + \psi(n))\lambda_2\big).$$

Now, since the two middle terms vanish, $d((\lambda_1, 0), (\lambda_2, 0)) = 0$ if and only if the terms $(\lambda_1, \delta_1 P'(\lambda)\lambda_2)$ and $(\lambda_1, \partial P'(n)\lambda_2)$ are as stated and $(\lambda_1, \partial P'(Z)\lambda_2) = 0$.

Proceeding similarly for the remaining cases yields the claim. ∎

Remarks 4.7. (a) The relationship between P' and P'_* is as follows. Fix $q_0 \in U'$ and write P' as

$$P'(p, u, q) = -l(p) + \pi(u) + Ad_u l(q) Ad_u^*$$

for some map $l : U' \to L(\mathfrak{g}'^*, \mathfrak{g}')$ with $l(q_0) = 0$ and some group cocycle $\pi : G' \to L(\mathfrak{g}'^*, \mathfrak{g}')$. We have

$$P'_*(q, \Lambda, X') = d\pi(1)\, X' + ad_{X'} l(q) + l(q) ad_{X'}^* + dl(q)\Lambda.$$

Therefore (this is a special case of a result of [**X**])

$$dl(q)(\Lambda) = P'_*(q, \Lambda, 0), \quad d\pi(u)T_1 l_u X' = Ad_u d\pi(1) X' Ad_u^* = Ad_u P'_*(q_0, 0, X') Ad_u^*.$$

(b) Writing out the equations of Theorem 4.6 for $\delta_1 P'$ using the basis (e_β) of $\mathfrak{n}$ and integrating yields

$$l(q)(e_\beta) = (\phi_\beta(q) - \phi_\beta(q_0))e_\beta, \text{ if } \beta \in <\Gamma>$$
$$l(q)(e_\beta) = e^{(\beta,\mu)}(e^{-(\beta,q_0)} - e^{-(\beta,q)})e_\beta, \text{ if } \beta \in \overline{\Gamma}^+$$
$$l(q)(e_\beta) = e^{-(\beta,\mu)}(e^{(\beta,q)} - e^{(\beta,q_0)})e_\beta, \text{ if } \beta \in \overline{\Gamma}^-$$
$$l(q)(\lambda) = (C^\#(q) - C^\#(q_0))(\lambda) \text{ for all } \lambda \in \mathfrak{h}^*$$

We will check the first line, the others are similar. Set $\psi(q)(e_\alpha) = \psi_\alpha(q)e_\alpha$. Then the first equation for $\delta_1 P'$ in Theorem 4.6 is equivalent to

$$(e_\alpha, dl(q)(\Lambda)e_\beta) = -(\Pi_\mathfrak{h}[\psi^* e_\alpha, \psi^* e_\beta], \Lambda)$$
$$= -\psi_\beta(q)\,\psi_{-\beta}(q)\,\delta_{\alpha,-\beta}\,(\beta, \Lambda),$$

while the second is $(\lambda, dl(q)(\Lambda)e_\beta) = 0$. Therefore, if $\beta \in <\Gamma>$,

$$dl(q)(\Lambda)e_\beta = -\psi_\beta(q)\psi_{-\beta}(q)(\beta, \Lambda)e_\beta$$
$$= \frac{-1}{4\sinh^2\left(\frac{(\beta,q-\mu)}{2}\right)}(\beta, \Lambda)\,e_\beta$$
$$= d\phi_\beta(q)\,(\Lambda)\,e_\beta.$$

On the other hand, the remaining equations evaluated at $q = q_0$ give

$$(\lambda, d\pi(1)(n)e_\beta) = -(\phi_\beta(q_0)(\lambda, \beta) + (C^\#(q_0)\lambda, \beta))(e_\beta, n)$$
$$(e_\alpha, d\pi(1)(n)e_\beta) = -\frac{\psi_{-\alpha}(q_0)\psi_{-\beta}(q_0)}{\psi_{-(\alpha+\beta)}(q_0)}([e_\alpha, e_\beta], n)$$
$$(\lambda, d\pi(1)(n)\lambda') = 0, \quad d\pi(1)Z = 0$$

for all $\alpha, \beta \in \Delta, \lambda, \lambda' \in \mathfrak{h}^*, Z + n \in \mathfrak{g}'$.

The latter equations allow, in principle, for an explicit expression of $\pi(u)$ but we will postpone this integration to a future publication as it will not be needed in the rest of this paper. (Note that from the expression for π_Σ in Remark 2.2.7 (c), we cannot remove the term π by a (base preserving) groupoid automorphism of X'.)

CHAPTER 5

Coboundary Dynamical Poisson Groupoids - The Constant R-Matrix Case

The purpose of this section is two-fold. In Section 5.1, we give a construction of the dual X^* of the coboundary dynamical Poisson groupoid $X = \mathfrak{h}^* \times G \times \mathfrak{h}^*$ (of Theorem 2.1.4) for the constant r-matrix case, i.e., for the case where R is a constant map from $\mathfrak{h}^*$ to $L(\mathfrak{g}^*, \mathfrak{g})$. As the reader will see, the construction involves the use of Poisson Lie group theory. More specifically, the Poisson Lie group G equipped with the Sklyanin bracket admits an extension to a bigger Poisson Lie group whose dual is critical in the construction. In Section 5.2, we construct a symplectic double groupoid which has X and X^* as its side groupoids. This leads, in particular, to a description of the symplectic leaves of X as orbits of a Poisson Lie group action. We will discuss the non-constant r-matrix case in a forthcoming publication.

5.1. The dual Poisson groupoid

Let $\iota : \mathfrak{h} \longrightarrow \mathfrak{g}$ be the inclusion map. We assume here that the Lie groups G and H are connected and simply connected. Let $R : \mathfrak{g}^* \longrightarrow \mathfrak{g}$ be a skew-symmetric constant r-matrix which satisfies (2.1.1) and (2.1.2). Recall that the group G equipped with the Sklyanin bracket

$$\{f, g\}_G(x) = <R(Df), Dg> - <R(D'f), D'g> \tag{5.1.1}$$

is a Poisson Lie group with tangent Lie bialgebra $(\mathfrak{g}, [\,,\,]; \mathfrak{g}^*, [\,,\,]_*)$ where

$$[A, B]_* = ad^*_{R(A)} B - ad^*_{R(B)} A. \tag{5.1.2}$$

Lemma 5.1.1. *H is a trivial Poisson Lie subgroup of G, that is, the Poisson structure on G vanishes on H.*

Proof. In a right invariant frame, the Hamiltonian operator associated with $\{\,,\,\}_G$ is given by $\eta(g) = R - Ad_g \circ R \circ Ad_g^*$. Since R is H-equivariant, the assertion is immediate from the expression for η. ∎

Let $(G^*, \{\,,\,\}_*)$ be Drinfel'd's Poisson Lie group dual to $(G, \{\,,\,\}_G)$, and let ([**STS**], [**LW**])

$$\varphi^+ : G^* \times \overline{G} \longrightarrow G^*, \quad \varphi^- : \overline{G^*} \times G \longrightarrow G \tag{5.1.3}$$

be the right and left dressing actions. Recall that φ^+ and φ^- are Poisson Lie group actions and that G and G^* act on each other by twisted automorphisms [**LW**]

$$\varphi^+_x(uv) = \varphi^+_{\varphi^-_v(x)}(u)\, \varphi^+_x(v), \quad \varphi^-_u(xy) = \varphi^-_u(x)\, \varphi^-_{\varphi^+_x(u)}(y). \tag{5.1.4}$$

They are related to (in fact defined by) the Poisson brackets of G and G^* by the formulas
$$\{\phi,\psi\}_*(u) = -d\phi(u)\lambda^+(T_1 l_u^* d\psi(u))(u)$$
$$\{f,g\}_G(x) = df(x)\lambda^-(T_1 r_x^* dg(x))(x), \tag{5.1.5}$$
where $\lambda^+(X)(u), \lambda^-(A)(x)$ are the infinitesimal generators of φ^+ and φ^-.

It follows from Lemma 5.1.1 that the restriction of the right dressing action to H induces a left Hamiltonian action
$$\phi^l : H \times G^* \longrightarrow G^* : (h,u) \mapsto \varphi^+_{h^{-1}}(u). \tag{5.1.6}$$

Moreover, since G^* acts trivially on $H \subset G$ (i.e. $\varphi^-_u(h) = h$, $u \in G^*$, $h \in H$), it is immediate from (5.1.4) that for each $h \in H$, ϕ^l_h is an automorphism of G^*.

Caveat Note that our conventions differ from those of [**LW1**]. Indeed φ^+ is their left dressing action made right, while φ^- is their right dressing action made left.

Recall also that the dressing vector fields λ^+ and λ^- may fail to globally integrate to define φ^+ and φ^-. In this subsection, we need only assume that ϕ^l is globally defined.

By construction, the map $\iota^* : \mathfrak{g}^* \longrightarrow \mathfrak{h}^*$ is a morphism of Lie algebras. Let
$$I^* : G^* \longrightarrow \mathfrak{h}^*$$
be the (unique) morphism of Lie groups integrating ι^*.

Lemma 5.1.2. *I^* is an Ad^*_H- equivariant momentum map for the action ϕ^l.*

Proof. For $Z \in \mathfrak{h}$, let $j_Z \in C^\infty(G^*, \mathfrak{h}^*)$ be defined by $j_Z(u) = <I^*(u), Z>$. We have to show that the Hamiltonian vector field $\widehat{X}_{j_Z}$ coincides with the infinitesimal generator $-\lambda^+(\iota Z)$ of the action ϕ^l. But
$$\frac{d}{dt}_{|t=0} j_Z(u e^{tA}) = \frac{d}{dt}_{|t=0} <I^*(u e^{tA}), Z>$$
$$= \frac{d}{dt}_{|t=0} <I^*(u) + I^*(e^{tA}), Z> = <\iota^* A, Z>.$$
Therefore
$$\widehat{X}_{j_Z}(u) = -\lambda^+(T_1 l_u^* dj_Z(u))(u)$$
$$= -\lambda^+(T_1 l_u^* T_u^* l_{u^{-1}} \iota Z)(u) = -\lambda^+(\iota Z)(u).$$
It remains to show that $Ad^*_{h^{-1}} I^*(u) = I^*(\varphi^+_{h^{-1}}(u))$. Since both sides are group morphisms from G^* to $\mathfrak{h}^*$ and G^* is connected, it is enough to check that the induced Lie morphisms are equal, that is,
$$\iota^* T_1 \varphi^+_{h^{-1}} A = Ad^*_{h^{-1}} \iota^* A.$$

But this equality follows from $T_1 \varphi^+_{h^{-1}} = Ad^*_{h^{-1}}$ and the Ad^*_H-equivariance of ι^*. Hence the claim. (Note that one may also use functoriality applied to the bialgebra morphism ι.) ∎

Proposition 5.1.3. *(a) The set $G \times \mathfrak{h}^*$ equipped with the multiplication $(x,p)(y,q) = (xy, p+q)$ and the Poisson bracket*

$$\{f,g\}(x,p) = <p, [\delta f, \delta g]> + <R(Df), Dg> - <R(D'f), D'g>$$
$$+ <(Df - D'f), \iota(\delta g)> - <(Dg - D'g), \iota(\delta f)>$$

is a Poisson Lie group.

(b) The Drinfel'd Poisson Lie group dual of $(G \times \mathfrak{h}^*, \{,\})$ is the set $H \times G^*$ equipped with the semi-direct multiplication

$$(h,u)(k,v) = (hk, u\varphi_{h^{-1}}^+(v))$$

and the Poisson bracket

$$\{\phi, \psi\}_*(h, u) = -\partial_* \phi \lambda^+ (T_1^* l_u \partial_* \psi)(u),$$

where $\partial_* \phi$ is the partial derivative with respect to G^*.

Proof. (a) For the Jacobi identity, we use the same notation J_{ijk} as in the proof of Theorem 2.2.5, with $i, j, k \in \{*, 1\}$ where, as an index, $*$ stands for G and 1 stands for $\mathfrak{h}^*$.

It is easy to see that $J_{111} = 0$ and $J_{*11} = 0$. Now, define $f_Z \in C^\infty(G \times \mathfrak{h}^*)$ by $f_Z(g,p) = <p, Z>$, let $\phi, \psi \in C^\infty(G)$ and denote by Y^l (resp. Y^r) the left (resp. right) invariant vector field on G with value $Y \in \mathfrak{g}$ at the identity. By direct calculation,

$$\{p_G^* \phi, \{p_G^* \psi, f_Z\}\} = \big((RD\phi)^r - (RD'\phi)^l\big)\big((\iota Z)^r - (\iota Z)^l\big)(\psi),$$

and

$$\{f_Z, \{p_G^* \phi, p_G^* \psi\}\}$$
$$= \big((\iota Z)^l - (\iota Z)^r\big)\big((RD\phi)^r - (RD'\phi)^l\big)(\psi) - (\phi \leftrightarrow \psi),$$

where $(R(D\phi))^l(g) = (R(D\phi(g)))^l(g)$, and similarly for the others. Collecting terms yields

$$J_{**1} = 0 \Leftrightarrow$$
$$<[D\phi, D\psi]_*, \iota Z> + <[D'\phi, D'\psi]_*, \iota Z> = 0, \quad \forall \phi, \psi \in C^\infty(G), \forall Z \in \mathfrak{h}.$$

But this holds by Lemma 5.1.1. Finally, the equality $J_{***} = 0$ is classical and is equivalent to the classical Yang-Baxter condition.

The verification that the multiplication map $(G \times \mathfrak{h}^*) \times (G \times \mathfrak{h}^*) \longrightarrow G \times \mathfrak{h}^*$ is Poisson is left to the reader.

(b) Computing the derivative $\frac{d}{dt}\big|_{t=0}\{f, g\}(e^{tY}, t\Lambda)$ shows that the tangent Lie bialgebra of $G \times \mathfrak{h}^*$ is given by $(\mathfrak{g} \oplus \mathfrak{h}^*, [,]_\oplus; \mathfrak{h} \ltimes \mathfrak{g}^*, [,]')$ where

$$[Z + A, Z' + A']' = [Z, Z'] - ad^*_{\iota(Z)} A' + ad^*_{\iota Z'} A + [A, A']_*.$$

Therefore the dual group is the semi-direct product $H \ltimes G^*$ as stated. Clearly the Poisson bracket $\{,\}_*$ linearizes to $\mathfrak{g} \oplus \mathfrak{h}^*$. Finally, the multiplicativity of $\{,\}_*$ is a consequence of (5.1.6) and the fact that G^* is a Poisson Lie group. This completes the proof.

Note that $\mathfrak{h} \ltimes \mathfrak{h}^\perp \subset \mathfrak{h} \ltimes \mathfrak{g}^*$ is a Lie subalgebra which is isomorphic to the vertex Lie algebra of Theorem 3.2.4. ∎

5.1. THE DUAL POISSON GROUPOID

Let $X = \mathfrak{h}^* \times G \times \mathfrak{h}^*$ be the dynamical Poisson groupoid of Theorem 2.1.4 with constant r-matrix taken to be $-R$. By the proof of Theorem 3.2.4 and Proposition 2.2.1, the dual groupoid X^* belongs to $\mathcal{C}_{\mathfrak{h}^*}$. In the theorem below we will give the explicit $\mathcal{C}_{\mathfrak{h}^*}$ structure of X^*.

If $f \in C^\infty(H \times \mathfrak{h}^* \times G^*)$ we define δf and $D'f$ by the formulas

$$<\delta f, \lambda> = \frac{d}{dt}_{|t=0} f(h, p + t\lambda, u), \quad <D'f, Z> = \frac{d}{dt}_{|t=0} f(he^{tZ}, p, u),$$

and we denote by $\partial_* f$ the partial derivative with respect to G^*.

Theorem 5.1.4. (Dual Poisson groupoid (second form)) *Let X be as above.*

(a) The set $\Gamma := H \times \mathfrak{h}^* \times G^*$ together with the product Poisson bracket

$$\{f, g\}_\Gamma(h, p, u) = - <D'g, \delta_1 f> + <D'f, \delta_1 g>$$
$$- <p, [\delta_1 f, \delta_1 g]> -\partial_* f \lambda^+ (T_1 l_u^* \partial_* g)(u),$$

the commuting Hamiltonian actions of H

$$\phi_k^-(h, p, u) = (kh, p, \varphi_{k^{-1}}^+ u), \quad \phi_k^+(h, p, u) = (hk, Ad_k^* p, u)$$

with equivariant momentum maps

$$j_-(h, p, u) = Ad_{h^{-1}}^* p + I^*(u), \quad j_+(h, p, u) = p$$

(I^* is as in Lemma 5.1.2), and the groupoid structure

$$\alpha = j_-, \beta = j_+, \epsilon(q) = (1, q, 1)$$
$$(h, j_-(k, q, v), u) \cdot (k, q, v) = (hk, q, u\varphi_{h^{-1}}^+ v)$$
$$i(h, p, u) = (h^{-1}, j_-(h, p, u), \varphi_h^+(u^{-1}))$$

is a Poisson groupoid in $\mathcal{C}_{\mathfrak{h}^*}$.

(b) The Poisson groupoid Γ of (a) is the Poisson groupoid dual X^* of X.

Proof. (a) That the actions $\phi^\pm$ are Hamiltonian with equivariant momentum maps $j_\pm$ follows from Example 2.1.3, the Hamiltonian property of the action ϕ^l ((5.1.6)), and Lemma 5.1.2. On the other hand, the automorphism property $\varphi_{h^{-1}}^+(uv) = \varphi_{h^{-1}}^+(u) \varphi_{h^{-1}}^+(v)$ and, again, the Ad_H^*- equivariance of the morphism I^*, imply that the groupoid axioms

$$m: \Gamma * \Gamma \longrightarrow \Gamma \text{ is associative}$$
$$\epsilon(\alpha(h, p, u)) \cdot (h, p, u) = (h, p, u) = (h, p, u) \cdot \epsilon(\beta(h, p, u))$$
$$(h, p, u) \cdot i(h, p, u) = \epsilon(\alpha(h, p, u)), \quad i(h, p, u) \cdot (h, p, u) = \epsilon(\beta(h, p, u))$$

are satisfied.

The lengthy check that the graph of the multiplication

$$Gr(m) \subset \Gamma \times \Gamma \times \overline{\Gamma}$$

is a coisotropic submanifold is postponed to the appendix.

(b) We have to show that the Lie bialgebroid tangent to Γ is isomorphic to the Lie bialgebroid $(A(X)^*, A(X)) \simeq (\mathfrak{h}^* \times \mathfrak{h} \times \mathfrak{g}^*, \mathfrak{h}^* \times \mathfrak{h}^* \times \mathfrak{g})$ of Proposition 3.2.1 (with constant r-matrix $-R$). We will only sketch the main steps.

(i) The isomorphism $A(\Gamma) \simeq A(X)^*$. We have to compute (see the end of section 2.1) the value on $\epsilon(\mathfrak{h}^*)$ of the Lie bracket of two left invariant sections

$$X^l, X'^l : \Gamma :\longrightarrow Ker\,T\alpha \subset T\Gamma.$$

We have

$$T_{(h,q,u)}j_-(Z(h), \rho, A(u)) = Ad^*_{h^{-1}}(-ad^*_{T_h l_{h^{-1}} Z(h)} q + \rho) + \iota^* T_u l_{u^{-1}} A(u),$$

where $Z(h) \in T_h H, \rho \in \mathfrak{h}^*, A(u) \in T_u G^*$. Therefore,

$$A(\Gamma)_q := Ker\,T_{(1,q,1)}j_- = \{(Z, ad^*_Z q - \iota^* A, A) \mid Z \in \mathfrak{h}, A \in \mathfrak{g}^*\} \simeq \mathfrak{h} \times \mathfrak{g}^*,$$

where the identification $\simeq$ is obtained by dropping the middle term. Let

$$L_{(h,q,u)} : j_-^{-1}(j_+(h, q, u)) \longrightarrow j_-^{-1}(j_-(h, q, u))$$
$$(k, Ad^*_k(q - I^*(v)), v) \mapsto (hk, Ad^*_k(q - I^*(v)), u\varphi^+_{h^{-1}} v)$$

be the left translation on Γ. The left invariant vector field X^l on Γ whose restriction to $\epsilon(\mathfrak{h}^*)$ (modulo $\simeq$) is

$$X : \mathfrak{h}^* \longrightarrow A(\Gamma) : q \mapsto (q, Z(q), A(q))$$

is given by

$$\begin{aligned}X^l(h, q, u) &= T_{\epsilon \circ j_+(h,q,u)} L_{(h,q,u)} X^l(\epsilon \circ j_+(h, q, u)) \\ &= T_{(1,q,1)} L_{(h,q,u)} (Z(q), ad^*_{Z(q)} q - \iota^* A(q), A(q)) \\ &= (T_1 l_h\, Z(q), ad^*_{Z(q)} q - \iota^* A(q), T_1 l_u Ad^*_{h^{-1}} A(q)).\end{aligned}$$

To calculate the commutator $[X^l, X'^l]_\Gamma(1, q, 1)$, we use the basic formula

$$X^l X'^l(f)(1, q, 1) = \frac{d}{ds}\Big|_{s=0} X'^l(f)(\gamma(s)) = \frac{d}{ds}\Big|_{s=0} \Big(\frac{d}{dt}\Big|_{t=0} f(\tilde{\gamma}(t))\Big),$$

where γ and $\tilde{\gamma}$ are arbitrary curves which satisfy $\gamma(0) = (1, q, 1)$, $\frac{d}{ds}\big|_{s=0} \gamma(s) = X^l(1, q, 1)$, and $\tilde{\gamma}(0) = \gamma(s)$, $\frac{d}{dt}\big|_{s=0} \tilde{\gamma}(t) = X'^l(\gamma(s))$. Choosing the curve

$$\tilde{\gamma}(t) = \big(e^{sZ(q)} e^{tZ'(q(s))}, Ad^*_{e^{tZ'(q(s))}}(q(s)) - \iota^* t A'(q(s)), e^{sA(q)} \varphi^+_{e^{-sZ(q)}}(e^{tA'(q(s))})\big),$$

where $q(s) = Ad^*_{e^{sZ(q)}} q - I^*(e^{sA(q)})$, a direct calculation shows that

$$[X, X'](q) := [X^l, X'^l]_\Gamma(1, q, 1)$$

is given, after the identification $\simeq$, by

$$\begin{aligned}[X, X'](q) = (&dZ'(ad^*_Z q - \iota^* A) - dZ(ad^*_{Z'} q - \iota^* A') + [Z(q), Z'(q)], \\ &+ dA'(ad^*_Z q - \iota^* A) - dA(ad^*_{Z'} q - \iota^* A') \\ &- ad^*_{\iota Z} A' + ad^*_{\iota Z'} A + [A(q), A'(q)]_*),\end{aligned}$$

where $(\mathfrak{g}^*, [\,,\,]_*)$ is as in (5.1.2) and all maps and differentials are evaluated at $q \in \mathfrak{h}^*$. Thus the bracket indeed coincides, up to sign, with the one given in Proposition 3.2.1 for $A(q) = \iota$, $K(q)Z = ad^*_Z q$, and P as in Remark 3.2.2 with r-matrix $-R$. Now, applying the Lie functor to the morphism

$$[\alpha, \beta] : H \times \mathfrak{h}^* \times G^* \to \mathfrak{h}^* \times \mathfrak{h}^* : (h, q, u) \mapsto (j_-(h, q, u), j_+(h, q, u))$$

shows that the anchor $a_*(q, Z, A) = (q, ad^*_Z q - \iota^* A)$.

(ii) The isomorphism $A(H \times \mathfrak{h}^* \times G^*)^* \simeq A(X)$. The unit section is $\epsilon : \mathfrak{h}^* \longrightarrow \Gamma : q \mapsto (1, q, 1)$. Therefore $\gamma \in N^*(\epsilon(\mathfrak{h}^*))_q$ if and only if $\gamma = (\lambda, 0, X)$ for some $\lambda \in \mathfrak{h}^*$ and $X \in \mathfrak{g}$.

Let $\theta, \theta' : \mathfrak{h}^* \longrightarrow N^*(\epsilon(\mathfrak{h}^*)) \simeq \mathfrak{h}^* \times \mathfrak{h}^* \times \mathfrak{g}$ be two sections expressed as $\theta(q) = (\lambda(q), 0, X(q))$ and $\theta'(q) = (\lambda'(q), 0, X'(q))$. We set

$$\overline{\theta}(h, q, u) = (T_h^* l_{h^{-1}} \lambda(q), 0, T_u^* l_{u^{-1}} X(q)),$$

and similarly for $\overline{\theta'}$. By (3.1.2), it suffices to calculate

$$[\theta, \theta'](q) = [\overline{\theta}, \overline{\theta'}](1, q, 1)$$
$$= (-L_{\Pi_\Gamma^\# \overline{\theta}} \overline{\theta'} + L_{\Pi_\Gamma^\# \overline{\theta'}} \overline{\theta} - d < \overline{\theta}, \Pi_\Gamma^\# \overline{\theta'} >)(1, q, 1),$$

where the Hamiltonian operator is given by

$$\Pi_\Gamma^\#(h, q, u)(\overline{\theta}) = (0, -\lambda(q), -\lambda^+(X(q))(u)).$$

Let Z^l (resp. A^l) be the left invariant vector field on H (resp. G^*) with $Z^l(1) = Z \in \mathfrak{h}$ (resp. $A^l(1) = A \in \mathfrak{g}^*$) and let $\rho \in \mathfrak{h}^*$.

We will calculate $< [\overline{\theta}, \overline{\theta'}], (Z^l, \rho, A^l) > (1, q, 1)$ making use of the formula

$$< L_{\Pi_\Gamma^\# \overline{\theta}} \overline{\theta'}, (Z^l, \rho, A^l) > = L_{\Pi_\Gamma^\# \overline{\theta}} < \overline{\theta'}, (Z^l, \rho, A^l) > + < \overline{\theta'}, [(Z^l, \rho, A^l), \Pi_\Gamma^\# \overline{\theta}]_\Gamma >$$
$$= I + II.$$

It is immediate to see that

$$I = -d\lambda'(q)(\lambda(q))Z - dX'(q)(\lambda(q))A.$$

As for II, let $\phi_t(h, q, u) = (he^{tZ}, q + t\rho, ue^{tA})$ be the flow of (Z^l, ρ, A^l). We have

$$II = < \overline{\theta'}, \frac{d}{dt}_{|t=0} T_{\phi_t(1,q,1)} \phi_{-t} \Pi_\Gamma^\# \overline{\theta}(\phi_t(1, q, 1)) >$$
$$= < \overline{\theta'}, \frac{d}{dt}_{|t=0} T_{\phi_t(1,q,1)} \phi_{-t}(0, -\lambda(q + t\rho), -\lambda^+(X(q + t\rho))(e^{tA})) >$$
$$= < \overline{\theta'}, \left(0, -d\lambda(q)(\rho), -\frac{d}{dt}_{|t=0} (T_{e^{tA}} r_{e^{-tA}} \lambda^+(X(q + t\rho))(e^{tA}))\right) >.$$

Now, making use of $\lambda^+(X)(u) = T_1 r_u \Pi_{\mathfrak{g}^*}(Ad_u^D X)$ where Ad_u^D is the adjoint action of Drinfel'd's double, we have

$$\frac{d}{dt}_{|t=0} (T_{e^{tA}} r_{e^{-tA}} \lambda^+(X(q + t\rho))(e^{tA})) = \frac{d}{dt}_{|t=0} \Pi_{\mathfrak{g}^*}(Ad_{e^{tA}}^D X(q + t\rho))$$
$$= \Pi_{\mathfrak{g}^*}(ad_A^D X(q) + dX(q)(\rho))$$
$$= ad^*_{X(q)} A.$$

Therefore,

$$II = < \overline{\theta'}, (0, -d\lambda(q)(\rho), -ad^*_{X(q)} A) >$$
$$= < (\lambda'(q), 0, X'(q)), (0, -d\lambda(q)(\rho), -ad^*_{X(q)} A) >$$
$$= - < [X(q), X'(q)], A >.$$

Proceeding similarly for $< d < \bar{\theta}, \Pi_\Gamma^\# \overline{\theta'} >, (Z^l, \rho, A^l) >$ and collecting terms shows that $[(\lambda, X), (\lambda', X')](q)$ coincides with that of the trivial Lie algebroid of Proposition 3.2.1. Finally observe that the anchor map, which is the restriction of $-\Pi^\#$ to $N^*(\epsilon(\mathfrak{h}^*))$, is given by $a(q, \lambda, X) = (q, \lambda)$. (Recall that the dressing field λ^+ vanishes at $u = 1$.) This concludes the proof of the theorem. ∎

In the special case when $R = 0$, we have $G^* = \mathfrak{g}^*$ equipped with the Lie-Poisson structure. In this case $I^* = \iota^* : \mathfrak{g}^* \longrightarrow \mathfrak{h}^*$ and $\varphi_{h^{-1}}^+(A) = Ad_{h^{-1}}^*(A)$. Specializing Theorem 5.1.4 to this situation, we have

Corollary 5.1.5. (Dual Poisson groupoid for $R = 0$.) *Let $X = \mathfrak{h}^* \times G \times \mathfrak{h}^*$ be the coboundary dynamical Poisson groupoid of Theorem 2.1.4 with $R = 0$. Then the Poisson groupoid dual X^* of X is the set $H \times \mathfrak{h}^* \times \mathfrak{g}^*$ equipped with the Poisson bracket*

$$\{f, g\}_*(h, p, A) = - < D'g, \delta_1 f > + < D'f, \delta_1 g >$$
$$- < p, [\delta_1 f, \delta_1 g] > + < A, [\delta f, \delta g] >,$$

$(< \delta f, B > = \frac{d}{dt}\big|_{t=0} f(h, p, A + tB))$, *and the groupoid structure*

$$\alpha(h, p, A) = Ad_{h^{-1}}^* p + \iota^* A, \quad \beta(h, p, A) = p, \quad \epsilon(p) = (1, p, 0)$$
$$(h, \alpha(k, q, B), A) \cdot (k, q, B) = (hk, q, A + Ad_{h^{-1}}^* B)$$
$$i(h, p, A) = (h^{-1}, \alpha(h, p, A), -Ad_h^* A). \quad \blacksquare$$

We now describe the trivialization of the Lie groupoid $\Gamma = H \times \mathfrak{h}^* \times G^*$.

Let $H^\perp$ be the connected and simply connected Lie subgroup of G^* with $Lie(H^\perp) = (\mathfrak{h}^\perp, [,]_*)$ and

$$H \ltimes H^\perp \subset H \ltimes G^*$$

be the Lie subgroup with Lie algebra $\mathfrak{h} \ltimes \mathfrak{h}^\perp \subset \mathfrak{h} \ltimes \mathfrak{g}^*$ (see the note at the end of the proof of Proposition 5.1.3).

Proposition 5.1.6. (Trivialization.) *Equip $\mathfrak{h}^* \times (H \ltimes H^\perp) \times \mathfrak{h}^*$ with the trivial Lie groupoid structure over $\mathfrak{h}^*$, and let Γ be the Lie groupoid in Theorem 5.1.4. If s is an arbitrary linear section $s : \mathfrak{h}^* \to \mathfrak{g}^*$ of ι^*, the map*

$$\Sigma : \mathfrak{h}^* \times (H \ltimes H^\perp) \times \mathfrak{h}^* \longrightarrow \Gamma$$
$$(p, (k, u), q) \mapsto (k, q, exp(s(p)) \, u \, \varphi_{k^{-1}}^+(exp(-s(q))))$$

is a Lie groupoid isomorphism.

Proof. We use an elementary device (see [**M1**]) according to which if $\sigma : \mathfrak{h}^* \longrightarrow \Gamma$ is a global smooth section of the restriction of α to the fiber $\beta^{-1}(0)$ and $G' \subset \Gamma$ is the isotropy subgroup at 0, then the map

$$\Sigma : \mathfrak{h}^* \times G' \times \mathfrak{h}^* \longrightarrow \Gamma$$
$$(p, x', q) \mapsto \sigma(p) \cdot x' \cdot i(\sigma(q))$$

is a Lie groupoid isomorphism.

Now $G' = \beta^{-1}(0) \cap \alpha^{-1}(0) = H \ltimes \mathfrak{H}^\perp$, while $\alpha|_{\beta^{-1}(0)} : \beta^{-1}(0) \to \mathfrak{h}^*$: $(h, 0, u) \mapsto I^*(u)$. Observe that for any linear section $s : \mathfrak{h}^* \to \mathfrak{g}^*$, the map

$$\sigma : \mathfrak{h}^* \to \Gamma : p \mapsto (1, 0, exp\, s(p))$$

is a smooth section of $\alpha|_{\beta^{-1}(0)}$ since $\alpha(\sigma(p)) = I^*(exp\, s(p)) = exp_{h^*} \iota^*(s(p)) = p$. Calculating $\sigma(p) \cdot (k, 0, u) \cdot i(\sigma(q))$ in Γ immediately yields the claim. ∎

Caveat Note that if $\mathfrak{h} \neq 0$, the group G' is not in general isomorphic to the Poisson Lie group dual G^*. For example, if $R = 0$, we have $G^* = \mathfrak{g}^*$ but $G' = H \ltimes \mathfrak{H}^\perp$. So G' and G^* may differ even topologically.

Remark 5.1.7. By Theorem 3.2.4, the Poisson bracket $\{\,,\,\}_*$ on $\mathfrak{h}^* \times (H \ltimes \mathfrak{H}^\perp) \times \mathfrak{h}^*$ defined by $\{\Sigma^* f, \Sigma^* g\}_* := \Sigma^* \{f, g\}_\Gamma$ has the form given by Theorem 2.2.5. However, even for standard Poisson Lie groups, the explicit expression of $\{\,,\,\}_*$ turns out to be rather cumbersome.

We close this subsection with the following

Example 5.1.8. Let $\mathfrak{g} = \mathfrak{h} \oplus \mathfrak{n}^+ \oplus \mathfrak{n}^-$ be the root space decomposition of a complex simple Lie algebra $\mathfrak{g}$, as in section 4. Let $R = \Pi_{\mathfrak{n}^+} - \Pi_{\mathfrak{n}^-}$ be the standard r-matrix. In what follows, we will scale the Poisson bracket by $1/2$ to match with standard conventions. Let $N^\pm$ be the (connected and simply connected) unipotent subgroups of G with Lie algebra $\mathfrak{n}^\pm$. Note that $H = \mathfrak{h} \simeq \mathfrak{h}^*$.

The dual group G^* is the set $\mathfrak{h} \times N^+ \times N^-$ with semi-direct group law

$$(Z, A, B) \cdot (Z', A', B') = (Z + Z', A e^{\frac{Z}{2}} A' e^{-\frac{Z}{2}}, e^{-\frac{Z}{2}} B' e^{\frac{Z}{2}} B).$$

The dressing action of $\mathfrak{h}$ on G^* is given by

$$\varphi^+_{-Y}(Z, A, B) = (Z, e^Y A e^{-Y}, e^Y B e^{-Y}).$$

The Poisson Lie group $H \ltimes G^*$ of Proposition 5.1.3 (b) is the set $\mathfrak{h} \times \mathfrak{h} \times N^+ \times N^-$ with group law

$$(Y, Z, A, B) \cdot (Y', Z', A', B') = (Y + Y', Z + Z', A e^{\frac{Z}{2}} e^Y A' e^{-Y} e^{-\frac{Z}{2}},$$
$$e^{-\frac{Z}{2}} e^Y B' e^{-Y} e^{\frac{Z}{2}} B),$$

and hence the vertex subgroup G' of the groupoid Γ is the set $\mathfrak{h} \times N^+ \times N^-$ with group law

$$(Y, A, B) \cdot (Y', A', B') = (Y + Y', A e^Y A' e^{-Y}, e^Y B' e^{-Y} B).$$

Finally, the map

$$\mathfrak{h}^* \times G' \times \mathfrak{h}^* \longrightarrow \Gamma = \mathfrak{h}^* \times \mathfrak{h} \times N^+ \times N^- \times \mathfrak{h}^*$$
$$(p, Y, A, B, q) \mapsto (q, Y, e^{\frac{p}{2}} A e^{-\frac{p}{2}}, e^{-\frac{p}{2}} B e^{\frac{p}{2}}, p - q)$$

gives an explicit trivialization of the Lie groupoid Γ.

5.2. Construction of the associated symplectic double groupoid

Our goal in this subsection is to construct, for the constant r-matrix case (taken to be $-R$), a symplectic double groupoid having $\mathfrak{h}^* \times G \times \mathfrak{h}^*$ and $H \times \mathfrak{h}^* \times G^*$ as its side Poisson groupoids.

We begin by recalling the notion of double Lie groupoids [E], [M3], and symplectic double groupoids [W1], [LW2], [M2].

Definition 5.2.1. *(a) A double Lie groupoid consists of a quadruple $(\mathcal{S}; \mathcal{H}, \mathcal{V}, B)$ where $\mathcal{H}$ and $\mathcal{V}$ are Lie groupoids over B, and $\mathcal{S}$ is equipped with two Lie groupoid structures, a horizontal structure with base $\mathcal{V}$, and a vertical structure with base $\mathcal{H}$, such that the structure maps (source, target, multiplication, unit section and inversion) of each groupoid structure on $\mathcal{S}$ are morphisms with respect to the other. We call $\mathcal{H}$ and $\mathcal{V}$ the side groupoids of $\mathcal{S}$, and B the double base. $(\mathcal{S}; \mathcal{H}, \mathcal{V}, B)$ is displayed as in Fig. 5.2.1 below.*

(b) A double Lie groupoid $(\mathcal{S}; \mathcal{H}, \mathcal{V}, B)$ is called symplectic if $\mathcal{S}$ is equipped with a symplectic structure such that both $\mathcal{S} \overset{\tilde{\alpha}_{\mathcal{H}}, \tilde{\beta}_{\mathcal{H}}}{\rightrightarrows} \mathcal{V}$ and $\mathcal{S} \overset{\tilde{\alpha}_{\mathcal{V}}, \tilde{\beta}_{\mathcal{V}}}{\rightrightarrows} \mathcal{V}$ are symplectic groupoids.

$$\begin{array}{ccc}
\mathcal{S} & \overset{\tilde{\alpha}_{\mathcal{H}}, \tilde{\beta}_{\mathcal{H}}}{\rightrightarrows} & \mathcal{V} \\
\tilde{\alpha}_{\mathcal{V}}, \tilde{\beta}_{\mathcal{V}} \downdownarrows & & \downdownarrows \alpha_{\mathcal{V}}, \beta_{\mathcal{V}} \\
\mathcal{H} & \overset{\alpha_{\mathcal{H}}, \beta_{\mathcal{H}}}{\rightrightarrows} & B
\end{array}$$

Fig. 5.2.1

We will consider the case where the Poisson Lie group G is complete. In this case, the Drinfel'd double D can be identified with $G \times G^*$ [STS], [LW1] with multiplication

$$(g_1, u_1) \cdot (g_2, u_2) = ((\varphi^-_{u_2^{-1}}(g_1^{-1}))^{-1} g_2, u_1 (\varphi^+_{g_1^{-1}}(u_2^{-1}))^{-1}). \qquad (5.2.2)$$

As a first step in the construction, we show that $X = \mathfrak{h}^* \times G \times \mathfrak{h}^*$ and $X^* = H \times \mathfrak{h}^* \times G^*$ form a matched pair of Lie groupoids in the sense of the following

Definition 5.2.2 [M3]. *Two Lie groupoids $\mathcal{V}$ and $\mathcal{H}$ over the same base B are said to form a matched pair of Lie groupoids iff the manifold*

$$\mathcal{V} * \mathcal{H} = \{(v, h) \in \mathcal{V} \times \mathcal{H} \mid \beta_{\mathcal{V}}(v) = \alpha_{\mathcal{H}}(h)\}$$

admits a Lie groupoid structure over B such that
(a) the maps $h \mapsto \overline{h} = (\epsilon_{\mathcal{V}}(\alpha_{\mathcal{H}}(h)), h)$ and $v \mapsto \overline{v} = (v, \epsilon_{\mathcal{H}}(\beta_{\mathcal{V}}(v)))$ are morphism of Lie groupoids from $\mathcal{H}$ and $\mathcal{V}$ to $\mathcal{V} * \mathcal{H}$ respectively,
(b) the map $\mathcal{V} * \mathcal{H} \longrightarrow \mathcal{V} * \mathcal{H} : (v, h) \mapsto \overline{v}\overline{h}$ is a diffeomorphism.
In this case, the groupoid $\mathcal{V} * \mathcal{H}$ is called the matched product of $\mathcal{V}$ and $\mathcal{H}$.

Proposition 5.2.3. *The Lie groupoids X and X^* form a matched pair with matched product given by the trivial groupoid $\mathfrak{h}^* \times M \times \mathfrak{h}^*$, where the vertex group $M = H \times (G \times G^*)$ is the direct product of H with the Drinfel'd double (see (5.2.2)).*

5.2. CONSTRUCTION OF THE ASSOCIATED SYMPLECTIC DOUBLE GROUPOID

Proof. Clearly, $X * X^*$ may be identified with the manifold $\mathfrak{h}^* \times M \times \mathfrak{h}^*$. Equip the latter with the trivial groupoid structure. The groupoids X and X^* are embedded as wide subgroupoids of $\mathfrak{h}^* \times M \times \mathfrak{h}^*$ through the morphisms $(p, g, q) \mapsto (p, 1, g, 1, q)$, $(h, p, u) \mapsto (Ad^*_{h^{-1}}p + I^*(u), h, h, u, p)$. Finally, for $(p, h, g, u, q) \in \mathfrak{h}^* \times M \times \mathfrak{h}^*$, we have the unique factorization

$$(p, h, g, u, q) = (p, 1, \varphi^-_u(gh^{-1}), 1, Ad^*_{h^{-1}}q + I^*(\varphi^+_{gh^{-1}}(u)))$$
$$\cdot (Ad^*_{h^{-1}}q + I^*(\varphi^+_{gh^{-1}}(u)), h, h, \varphi^+_{gh^{-1}}(u), q).$$

Hence it follows that (X, X^*) is a matched pair. ∎

We will denote by $\overline{X}$ and $\overline{X}^*$ the images of X and X^* under the morphisms in the proof above.

Given $(h, p, u) \in X^*, (p, g, q) \in X$, the corresponding elements in $\mathfrak{h}^* \times M \times \mathfrak{h}^*$ are composable, and we have the unique factorization

$$(Ad^*_{h^{-1}}p + I^*(u), h, h, u, q)(p, 1, g, 1, q)$$
$$= \left(Ad^*_{h^{-1}}p + I^*(u), 1, \varphi^-_u(hgh^{-1}), 1, Ad^*_{h^{-1}}q + I^*(\varphi^+_{hgh^{-1}}(u))\right) \quad (5.2.3)$$
$$\left(Ad^*_{h^{-1}}q + I^*(\varphi^+_{hgh^{-1}}(u)), h, h, \varphi^+_{hgh^{-1}}(u), q\right).$$

We therefore obtain two maps

$$\psi^- : (H \times \mathfrak{h}^* \times G^*) *_{\alpha_X} (\mathfrak{h}^* \times G \times \mathfrak{h}^*) \longrightarrow \mathfrak{h}^* \times G \times \mathfrak{h}^* \quad (5.2.4.a)$$
$$((h, p, u), (p, g, q)) \mapsto \left(Ad^*_{h^{-1}}p + I^*(u), \varphi^-_u(hgh^{-1}), Ad^*_{h^{-1}}q + I^*(\varphi^+_{hgh^{-1}}(u))\right),$$

and

$$\psi^+ : (H \times \mathfrak{h}^* \times G^*) *_{\beta_{X^*}} (\mathfrak{h}^* \times G \times \mathfrak{h}^*) \longrightarrow H \times \mathfrak{h}^* \times G^* \quad (5.2.4.b)$$
$$((h, p, u), (p, g, q)) \mapsto \left(h, q, \varphi^+_{hgh^{-1}}(u)\right).$$

Proposition 5.2.4. *ψ^- is a left groupoid action of X^* on X with moment map α_X and ψ^+ is a right groupoid action of X on X^* with moment map β_{X^*}. Furthermore, the following conditions are satisfied*

(a) $\beta_X(\psi^-_{g_-}(g_+)) = \alpha_{X^*}(\psi^+_{g_+}(g_-))$,

for all $g_+ \in X, g_- \in X^*$ with $\beta_{X^*}(g_-) = \alpha_X(g_+)$,

(b) $\psi^-_{g_-}(g_+ g'_+) = \psi^-_{g_-}(g_+)\psi^-_{\psi^+_{g_+}(g_-)}(g'_+)$,

for all $g_- \in X^*, g_+, g'_+ \in X$ with $\beta_{X^*}(g_-) = \alpha_X(g_+), \beta_X(g_+) = \alpha_X(g'_+)$,

(c) $\psi^+_{g_+}(g_- g'_-) = \psi^+_{\psi^-_{g'_-}(g_+)}(g_-)\psi^+_{g_+}(g'_-)$,

for all $g_+ \in X, g_-, g'_- \in X^*$ with $\beta_{X^*}(g_-) = \alpha_{X^*}(g'_-), \beta_{X^*}(g'_-) = \alpha_X(g_+)$.

Proof. The proof consists of direct checking and we will omit the details. See, however, Proposition 5.2.9 below. ∎

From standard consideration [**M3**], the upshot of the above proposition is that one can construct a vacant double Lie groupoid $(\mathcal{S}_{vac}; X^*, X, \mathfrak{h}^*)$ (vacant means that the double source map $\beta_+ : \mathcal{S}_{vac} \to X^* *_\beta X$ is a diffeomorphism) having X and X^* as its side groupoids. Indeed the horizontal structure of the vacant double is given by the left action groupoid $X^* \ltimes X$ corresponding to ψ^-, while the vertical structure is given by the right action groupoid $X^* \rtimes X$ associated with ψ^+. However, $\mathcal{S}_{vac}$ is not the correct underlying double Lie groupoid of the symplectic double groupoid which we are looking for, as is clear from dimension considerations. Nevertheless, as we will show in what follows, the sought for double Lie groupoid can be constructed by extending the objects X and X^*. It turns out that these extended objects act on the unextended ones through groupoid actions which restrict to $\psi^\pm$, and the corresponding left and right action groupoids then give the desired horizontal and vertical structures respectively. Before we carry out the details of this construction, let us make an important remark. As we know from Theorem 5.1.4, the Poisson structure on X^* is the product of the standard symplectic structure on $H \times \mathfrak{h}^*$ and the multiplicative structure on G^*. Since $H \times \mathfrak{h}^*$ is symplectic, the coarse groupoid $(H \times \mathfrak{h}^*) \times (H \times \mathfrak{h}^*)^- \rightrightarrows H \times \mathfrak{h}^*$ is a symplectic groupoid. On the other hand, there is a symplectic groupoid $G \times G^* \rightrightarrows G^*$ [**Lu**] with structure maps given as follows:

$$\alpha(g, u) = \varphi^+_{g^{-1}}(u^{-1})^{-1}, \quad \beta(g, u) = u$$
$$(g, \alpha(g', u')) \cdot (g', u') = (gg', u'). \quad (5.2.5)$$

Therefore the product groupoid

$$(H \times \mathfrak{h}^*) \times (H \times \mathfrak{h}^*)^- \times (G \times G^*) \rightrightarrows H \times \mathfrak{h}^* \times G^* \quad (5.2.6)$$

is a symplectic groupoid over X^*. It turns out that this product groupoid is isomorphic to the right action groupoid alluded to above.

We now introduce the extensions of X and X^*. Since H is a groupoid over a point, we have the product groupoid

$$X^*_e = X^* \times H \rightrightarrows \mathfrak{h}^* \quad (5.2.7).$$

Let $J_+ = \alpha_X$ and define $\Psi^- : X^*_e *_{J_+} X \longrightarrow X$ by

$$\Psi^-_{(h,p,u,k)}(p,g,q) = \left(Ad^*_{h^{-1}}p + I^*(u), \varphi^-_u(hgk^{-1}), Ad^*_{k^{-1}}q + I^*(\varphi^+_{hgk^{-1}}(u))\right). \quad (5.2.8)$$

Proposition 5.2.5. *Ψ^- is a left groupoid action of X^*_e on X with moment map J_+ such that $\Psi^-_{(h,p,u,h)}(p,g,q) = \psi^-_{(h,p,u)}(p,g,q)$ for all $(h,p,u) \in X^*$, $(p,g,q) \in X$.*

Proof. Clearly, $(\tilde{x}, x) = ((\tilde{h}, Ad^*_{h^{-1}}p + I^*(u), \tilde{u}, \tilde{k}), (h,p,u,k))$ is a composable pair in X^*_e and we have

$$\tilde{x}x = (\tilde{h}h, p, \tilde{u}\varphi^+_{\tilde{h}^{-1}}(u), \tilde{k}k).$$

Therefore,

$$\Psi^-_{\tilde{x}x}(p,g,q) = \left(Ad^*_{(\tilde{h}h)^{-1}}p + I^*(\tilde{u}\varphi^+_{\tilde{h}^{-1}}(u)), \varphi^-_{\tilde{u}\varphi^+_{\tilde{h}^{-1}}(u)}(\tilde{h}hg(\tilde{k}k)^{-1}),\right.$$
$$\left. Ad^*_{(\tilde{k}k)^{-1}}q + I^*(\varphi^+_{\tilde{h}hg(\tilde{k}k)^{-1}}(\tilde{u}\varphi^+_{\tilde{h}^{-1}}(u)))\right).$$

5.2. CONSTRUCTION OF THE ASSOCIATED SYMPLECTIC DOUBLE GROUPOID

On the other hand,

$$\Psi_{\tilde{x}}^{-}\Psi_{x}^{-}(p,g,q) = \big(Ad^*_{(\tilde{h}h)^{-1}}p + Ad^*_{\tilde{h}^{-1}}I^*(u) + I^*(\tilde{u}),\, \varphi^-_{\tilde{u}}(\tilde{h}\varphi^-_u(hgk^{-1})\tilde{k}^{-1}),$$
$$Ad^*_{(\tilde{k}k)^{-1}}q + Ad^*_{\tilde{k}^{-1}}I^*(\varphi^+_{hgk^{-1}}(u)) + I^*(\varphi^+_{\tilde{h}\varphi^-_u(hgk^{-1})\tilde{k}^{-1}}(\tilde{u})\big).$$

Since I^* is an H-equivariant homomorphism, the equality of the first components is clear. Now using the same property of I^*, we have

$$Ad^*_{\tilde{k}^{-1}}I^*(\varphi^+_{hgk^{-1}}(u))) + I^*(\varphi^+_{\tilde{h}\varphi^-_u(hgk^{-1})\tilde{k}^{-1}}(\tilde{u}))$$
$$= I^*(\varphi^+_{\tilde{h}\varphi^-_u(hgk^{-1})\tilde{k}^{-1}}(\tilde{u})\,\varphi^+_{\tilde{h}hg(\tilde{k}k)^{-1}}(\varphi^+_{\tilde{h}^{-1}}(u))).$$

But from (5.1.4) and the fact that H is a trivial Poisson Lie subgroup of G, we find

$$\varphi^+_{\tilde{h}hg(\tilde{k}k)^{-1}}(\tilde{u}\varphi^+_{\tilde{h}^{-1}}(u)) = \varphi^+_{\varphi^+_{\tilde{h}^{-1}}(u)}(\tilde{h}hg(\tilde{k}k)^{-1})(\tilde{u})\,\varphi^+_{\tilde{h}hg(\tilde{k}k)^{-1}}(\varphi^+_{\tilde{h}^{-1}}(u))$$
$$= \varphi^+_{\tilde{h}\varphi^-_u(hg(\tilde{k}k)^{-1})}(\tilde{u})\,\varphi^+_{\tilde{h}h(\tilde{k}k)^{-1}}(\varphi^+_{\tilde{h}^{-1}}(u))$$
$$= \varphi^+_{\tilde{h}\varphi^-_u(hgk^{-1})\tilde{k}^{-1}}(\tilde{u})\,\varphi^+_{\tilde{h}hg(\tilde{k}k)^{-1}}(\varphi^+_{\tilde{h}^{-1}}(u)).$$

Hence we have equality of the third components. Finally it follows from the above calculation that

$$\varphi^-_{\tilde{u}}(\tilde{h}\varphi^-_u(hgk^{-1})\tilde{k}^{-1}) = \varphi^-_{\tilde{u}}\varphi^-_{\varphi^+_{\tilde{h}^{-1}}(u)}(\tilde{h}hg(\tilde{k}k)^{-1})$$
$$= \varphi^-_{\tilde{u}\varphi^+_{\tilde{h}^{-1}}(u)}(\tilde{h}hg(\tilde{k}k)^{-1}).$$

This completes the proof that $\Psi_{\tilde{x}x}^{-} = \Psi_{\tilde{x}}^{-}\Psi_{x}^{-}$. The assertion on the relationship between Ψ^- and ψ^- is clear. ∎

The following corollary is a direct consequence of the definition of an action groupoid.

Corollary 5.2.6. *The left action groupoid $X_e^* \ltimes X \rightrightarrows X$ corresponding to Ψ^- has structure maps given by*

$$\tilde{\alpha}_\mathcal{H} : X_e^* \ltimes X \to X : ((h,p,u,k),(p,g,q)) \mapsto \Psi^-_{(h,p,u,k)}(p,g,q)$$
$$\tilde{\beta}_\mathcal{H} : X_e^* \ltimes X \to X : ((h,p,u,k),(p,g,q)) \mapsto (p,g,q)$$
$$\tilde{m}_\mathcal{H} : (X_e^* \ltimes X) * (X_e^* \ltimes X) \to X_e^* \ltimes X$$
$$((h_1,p_1,u_1,k_1),\Psi^-_{(h_2,p_2,u_2,k_2)}(p_2,g_2,q_2)) \cdot ((h_2,p_2,u_2,k_2),(p_2,g_2,q_2))$$
$$= ((h_1 h_2, p_2, u_1 \varphi^+_{h_1^{-1}}(u_2), k_1 k_2),(p_2,g_2,q_2))$$
$$\tilde{\epsilon}_\mathcal{H} : X \to X_e^* \ltimes X : (p,g,q) \mapsto ((1,p,1,1),(p,g,q))$$
$$\tilde{i}_\mathcal{H} : X_e^* \ltimes X \to X_e^* \ltimes X : ((h,p,u,k),(p,g,q)) \mapsto$$
$$((h^{-1}, Ad^*_{h^{-1}}p + I^*(u), \varphi^+_h(u^{-1}), k^{-1}),$$
$$(Ad^*_{h^{-1}}p + I^*(u), \varphi^-_u(hgk^{-1}), Ad^*_{k^{-1}}p + I^*(\varphi^+_{hgk^{-1}}(u)))). \blacksquare$$

For the extension of X, we consider the coarse groupoid $H \times H \rightrightarrows H$ and let

$$X_e = (H \times H) \times X \rightrightarrows H \times \mathfrak{h}^* \tag{5.2.9}$$

be the product groupoid. Introduce the map
$$J_- : X^* \longrightarrow H \times \mathfrak{h}^* : (h,p,u) \mapsto (h,p) \tag{5.2.10}$$
and define
$$\Psi^+ : X^* *_{J_-} X_e \longrightarrow X^* : \Psi^+_{(h,k,p,g,q)}(h,p,u) = (k,q,\varphi^+_{hgk^{-1}}(u)). \tag{5.2.11}$$

Proposition 5.2.7. Ψ^+ *is a right groupoid action of X_e on X^* with moment map J_- and we have $\Psi^+_{(h,h,p,g,q)}(h,p,u) = \psi^+_{(p,g,q)}(h,p,u)$ for all $(h,p,u) \in X^*, (p,g,q) \in X$.*

Proof. From the definition of J_- and Ψ^+, we have
$$J_-(\Psi^+_{(h,k,p,g,q)}(h,p,u)) = J_-(k,q,\varphi^+_{hgk^{-1}}(u))$$
$$= (k,q)$$
$$= \beta_{X_e}(h,k,p,g,q).$$
Consider the composable pair $(y,\tilde{y}) = \big((h,k,p,g,q),(k,l,q,\tilde{g},r)\big)$ in X_e. We have
$$\Psi^+_{y\tilde{y}}(h,p,u) = \Psi^+_{(h,l,p,g\tilde{g},r)}(h,p,u)$$
$$= (l,r,\varphi^+_{hg\tilde{g}l^{-1}}(u)).$$
Since $\varphi^+_{hg\tilde{g}l^{-1}} = \varphi^+_{k\tilde{g}l^{-1}}\varphi^+_{hgk^{-1}}(u)$, it follows that
$$\Psi^+_{y\tilde{y}}(h,p,u) = \Psi^+_{\tilde{y}}(k,q,\varphi^+_{hgk^{-1}}(u))$$
$$= \Psi^+_{\tilde{y}}\Psi^+_y(h,p,u).$$
On the other hand,
$$\Psi^+_{\epsilon_{X_e}(J_-(h,p,u))}(h,p,u) = \Psi^+_{(h,h,p,1,p)}(h,p,u) = (h,p,u).$$
Lastly, the assertion on the relationship between Ψ^+ and ψ^+ is clear from the definitions of the maps. ∎

Corollary 5.2.8. *The right action groupoid $X^* \rtimes X_e \rightrightarrows X^*$ corresponding to Ψ^+ has structure maps given by*

$$\tilde{\alpha}_\mathcal{V} : X^* \rtimes X_e \to X^* : \big((h,p,u),(h,k,p,g,q)\big) \mapsto (h,p,u)$$
$$\tilde{\beta}_\mathcal{V} : X^* \rtimes X \to X^* : \big((h,p,u),(h,k,p,g,q)\big) \mapsto \Psi^+_{(h,k,p,g,q)}(h,p,u)$$
$$\tilde{m}_\mathcal{V} : (X^* \rtimes X_e) * (X^* \rtimes X_e) \to X^* \rtimes X_e :$$
$$\big((h_1,p_1,u_1),(h_1,k_1,p_1,g_1,q_1)\big) \cdot \big(\Psi^+_{(h_1,k_1,p_1,g_1,q_1)}(h_1,p_1,u_1),(k_1,k_2,q_1,g_2,q_2)\big)$$
$$= \big((h_1,p_1,u_1),(h_1,k_2,p_1,g_1g_2,q_2)\big)$$
$$\tilde{\epsilon}_\mathcal{V} : X^* \to X^* \rtimes X_e : (h,p,u) \mapsto \big((h,p,u),(h,h,p,1,p)\big)$$
$$\tilde{i}_\mathcal{V} : X^* \rtimes X_e \to X^* \rtimes X_e : \big((h,p,u),(h,k,p,g,q)\big) \mapsto$$
$$\big((k,q,\varphi^+_{(hgk^{-1})}(u)),(h,h,q,g^{-1},p)\big). \blacksquare$$

Let $Pr_1 : X^* \longrightarrow H$ be the projection onto the first factor of $X^* = H \times \mathfrak{h}^* \times G^*$.

5.2. CONSTRUCTION OF THE ASSOCIATED SYMPLECTIC DOUBLE GROUPOID

Proposition 5.2.9. *The groupoid actions $\Psi^\pm$ satisfy the following properties:*

(a) $\beta_X(\Psi^-_{(g_-,k)}(g_+)) = \alpha_{X^*}(\Psi^+_{(Pr_1(g_-),k,g_+)}(g_-))$

for all $g_- \in X^*, g_+ \in X$ with $\beta_{X^*}(g_-) = \alpha_X(g_+)$ and all $k \in H$.

(b) $\Psi^-_{(g_-,k')}(g_+ g'_+) = \Psi^-_{(g_-,k)}(g_+) \Psi^-_{(\Psi^+_{(Pr_1(g_-),k,g_+)}(g_-),k')}(g'_+)$

for all $g_- \in X^*, g_+, g'_+ \in X$ with $\beta_{X^*}(g_-) = \alpha_X(g_+), \beta_X(g_+) = \alpha_X(g'_+)$ and for all $k, k' \in H$.

(c)
$$\Psi^+_{(Pr_1(g_-)Pr_1(g'_-),kk',g_+)}(g_- g'_-) =$$
$$\Psi^+_{(Pr_1(g_-),k,\Psi^-_{(g'_-,k')}(g_+))}(g_-) \, \Psi^+_{(Pr_1(g'_-),k',g_+)}(g'_-)$$

for all $g_+ \in X$, $g_-, g'_- \in X^*$ with $\beta_{X^*}(g_-) = \alpha_{X^*}(g'_-)$, $\beta_{X^*}(g'_-) = \alpha_X(g_+)$, and for all $k, k' \in H$.

Proof. We will check (b) and (c).

(b) Let $g_- = (h,p,u)$, $g_+ = (p,g_1,q)$, $g'_+ = (q,g_2,r)$. Then

$$\Psi^-_{(g_-,k')}(g_+,g'_+)$$
$$= \left(Ad^*_{h^{-1}}p + I^*(u), \varphi^-_u(hg_1g_2k'^{-1}), Ad^*_{k'^{-1}}r + I^*(\varphi^+_{hg_1g_2k'^{-1}}(u))\right),$$

and

$$\Psi^-_{(g_-,k)} \Psi^-_{(\Psi^+_{(Pr_1(g_-),k,g_+)}(g_-),k')}(g'_+) = \left(Ad^*_{h^{-1}}p + I^*(u),\right.$$
$$\left.\varphi^-_u(hg_1k^{-1})\varphi^-_{\varphi^+_{hg_1k^{-1}}(u)}(kg_2k'^{-1}), Ad^*_{k'^{-1}}r + I^*(\varphi^+_{hg_1g_2k'^{-1}}(u))\right).$$

Hence the assertion follows from (5.1.4).

(c) Let $g_- = (h,p,u)$, $g'_- = (h',p',u')$ satisfy $\beta_{X^*}(g_-) = \alpha_{X^*}(g'_-)$, i.e. $p = Ad^*_{h'^{-1}}p' + I^*(u')$, and let $g_+ = (p',g,q)$ so that $\beta_{X^*}(g'_-) = \alpha_X(g_+)$. We have

$$\Psi^+_{(Pr_1(g_-)Pr_1(g'_-),kk',g_+)}(g_- g'_-)$$
$$= (kk', q, \varphi^+_{hh'g(kk')^{-1}}(u\varphi^+_{h^{-1}}(u')))$$
$$= (kk', q, \varphi^+_{\varphi^-_{h^{-1}}(u')}(hh'g(kk')^{-1})}(u) \, \varphi^+_{h'g(kk')^{-1}}(u')).$$

On the other hand,

$$\Psi^+_{(Pr_1(g_-),k,\Psi^-_{(g'_-,k')}(g_+))}(g_-) \, \Psi^+_{(Pr_1(g'_-),k',g_+)}(g'_-)$$
$$= (kk', q, \varphi^+_{h\varphi^-_{u'}(h'gk'^{-1})k^{-1}}(u) \, \varphi^+_{h'g(kk')^{-1}}(u')).$$

The assertion then follows from the calculation in the proof of Proposition 5.2.5. ∎

Let $\mathcal{S} = X^*_e *_{J_+} X$. Then $\mathcal{S}$ supports both the left action groupoid structure $X^*_e \ltimes X$ and the right action groupoid structure $X^* \rtimes X_e$.

48   5. THE CONSTANT R-MATRIX CASE

Theorem 5.2.10. *If the horizontal structure on $\mathcal{S}$ is $X_e^* \bowtie X$ and the vertical structure on $\mathcal{S}$ is $X^* \bowtie X_e$, then $(\mathcal{S}; X^*, X, \mathfrak{h}^*)$ is a double Lie groupoid.*

Proof. We have to show that the structure maps of the horizontal (resp. vertical) structure on $\mathcal{S}$ are morphisms with respect to the vertical (resp. horizontal) structure. We will illustrate the role played by the properties in Proposition 5.2.9 by checking that $\tilde{\alpha}_{\mathcal{H}}$, $\tilde{\beta}_{\mathcal{V}}$, $\tilde{i}_{\mathcal{H}}$, and $\tilde{i}_{\mathcal{V}}$ are groupoid morphisms.

(i) $\tilde{\alpha}_{\mathcal{H}}$ is a groupoid morphism.

$$\tilde{\alpha}_{\mathcal{H}}\big((g_-, k, g_+) \cdot (\Psi^+_{(Pr_1(g_-), k, g_+)}(g_-), k', g'_+)\big)$$
$$= \tilde{\alpha}_{\mathcal{H}}(g_-, k', g_+ g'_+)$$
$$= \Psi^-_{(g_-, k')}(g_+ g'_+)$$
$$= \Psi^-_{(g_-, k)}(g_+) \Psi^-_{\Psi^+_{(Pr_1(g_-), k, g_+)}(g_-), k')}(g'_+) \text{ by Proposition 5.2.9 (b)}$$
$$= \tilde{\alpha}_{\mathcal{H}}(g_-, k, g_+) \, \tilde{\alpha}_{\mathcal{H}}(\Psi^+_{(Pr_1(g_-), k, g_+)}(g_-), k', g'_+).$$

(ii) $\tilde{\beta}_{\mathcal{V}}$ is a groupoid morphism.

$$\tilde{\beta}_{\mathcal{V}}\big((g_-, k, \Psi^-_{(g'_-, k')}(g'_+)) \cdot (g'_-, k', g'_+)\big)$$
$$= \tilde{\beta}_{\mathcal{V}}(g_- g'_-, kk', g'_+)$$
$$= \Psi^+_{(Pr_1(g_- g'_-), kk', g'_+)}(g_- g'_-)$$
$$= \Psi^+_{(Pr_1(g_-), k, \Psi^-_{(g'_-, h')}(g'_+))}(g_-) \, \Psi^+_{(Pr_1(g'_-), k', g'_+)}(g'_-)$$

by Proposition 5.2.9 (c)

$$= \tilde{\beta}_{\mathcal{V}}(g_-, k, \Psi^-_{(g'_-, k')}(g'_+)) \tilde{\beta}_{\mathcal{V}}(g'_-, k', g'_+).$$

(iii) $\tilde{i}_{\mathcal{H}}$ is a groupoid morphism.

$$\tilde{i}_{\mathcal{H}}\big((g_-, k, g_+) \cdot (\Psi^+_{(Pr_1(g_-), k, g_+)}(g_-), k', g'_+)\big)$$
$$= \tilde{i}_{\mathcal{H}}(g_-, k', g_+ g'_+)$$
$$= (g_-^{-1}, k'^{-1}, \Psi^-_{(g_-, k')}(g_+ g'_+)).$$

On the other hand,

$$\tilde{i}_{\mathcal{H}}(g_-,k,g_+)\,\tilde{i}_{\mathcal{H}}(\Psi^+_{(Pr_1(g_-),k,g_+)}(g_-),k',g'_+)$$
$$= (g_-^{-1},k^{-1},\Psi^-_{(g_-,k)}(g_+))((\Psi^+_{(Pr_1(g_-),k,g_+)}(g_-))^{-1},k'^{-1},$$
$$\Psi^-_{(\Psi^+_{(Pr_1(g_-),k,g_+)}(g_-),k')}(g'_+))$$
$$= (g_-^{-1},k^{-1},\Psi^-_{(g_-,h)}(g_+))(\Psi^+_{(Pr_1(g_-^{-1}),k^{-1},\Psi^-_{(g_-,k)}(g_+))}(g_-^{-1}),k'^{-1},$$
$$\Psi^-_{(\Psi^+_{(Pr_1(g_-),k,g_+)}(g_-),k')}(g'_+))$$

by Proposition 5.2.9 (c)

$$= (g_-^{-1},k'^{-1},\Psi^-_{(g_-,k)}(g_+)\Psi^+_{(\Psi^+_{(Pr_1(g_-),k,g_+)}(g_-),k')}(g'_+)).$$

Hence the assertion now follows from Proposition 5.2.9 (b).

(iv) $\tilde{i}_{\mathcal{V}}$ is a groupoid morphism.

$$\tilde{i}_{\mathcal{V}}\big((g_-,k,\Psi^-_{(g'_-,k')}(g'_+))\cdot(g'_-,k',g'_+)\big)$$
$$= \tilde{i}_{\mathcal{V}}(g_-g'_-,kk',g'_+)$$
$$= (\Psi^+_{(Pr_1(g_-g'_-),kk',g'_+)}(g_-g'_-),Pr_1(g_-g'_-),g'^{-1}_+)$$
$$= (\Psi^+_{(Pr_1(g_-),k,\Psi^-_{(g'_-,k')}(g'_+))}(g_-)\,\Psi^+_{(Pr_1(g'_-),k',g_+)}(g'_-),Pr_1(g_-)Pr_1(g'_-),g'^{-1}_-),$$

by Theorem 5.1.4 and Proposition 5.2.9 (c)

$$= (\Psi^-_{(Pr_1(g_-),k,\Psi^-_{(g'_-,k')}(g'+))}(g_-),Pr_1(g'_-),\Psi^-_{(g'_-,k')}(g'_-)^{-1})\cdot$$
$$(\Psi^+_{(Pr_1(g'_-),k',g'_+)}(g'_-),Pr_1(g'_-),g'^{-1}_+)$$
$$= \tilde{i}_{\mathcal{V}}(g_-,k,\Psi^-_{(g'_-,k')}(g'_+))\,\tilde{i}_{\mathcal{V}}(g'_-,k',g'_+).$$

We will leave the rest of the proof to the interested reader. ∎

To clarify the relation between the vacant double Lie groupoid $(\mathcal{S}_{vac};X^*,X,\mathfrak{h}^*)$ and the double Lie groupoid $(\mathcal{S};X^*,X,\mathfrak{h}^*)$ in the above theorem, we introduce the following definition

Definition 5.2.11. Let $(\mathcal{S}_1;\mathcal{H}_1,\mathcal{V}_1,P_1)$ be a double Lie groupoid. A double Lie subgroupoid of $(\mathcal{S}_1;\mathcal{H}_1,\mathcal{V}_1,P_1)$ is a double Lie groupoid $(\mathcal{S}_2;\mathcal{H}_2,\mathcal{V}_2,P_2)$ such that the Lie groupoids $\mathcal{S}_2 \rightrightarrows \mathcal{H}_2$, $\mathcal{S}_2 \rightrightarrows \mathcal{V}_2$, $\mathcal{H}_2 \rightrightarrows P_2$, $\mathcal{V}_2 \rightrightarrows P_2$ are respectively Lie subgroupoids of $\mathcal{S}_1 \rightrightarrows \mathcal{H}_1$, $\mathcal{S}_1 \rightrightarrows \mathcal{V}_1$, $\mathcal{H}_1 \rightrightarrows P_1$, $\mathcal{V}_1 \rightrightarrows P_1$.

Corollary 5.2.12. The vacant double Lie groupoid $(\mathcal{S}_{vac};X^*,X,\mathfrak{h}^*)$ associated with the matched pair (X,X^*) is a double Lie subgroupoid of the double Lie groupoid $(\mathcal{S};X^*,X,\mathfrak{h}^*)$ in Theorem 5.2.10.

Proof. As the side groupoids of the two double Lie groupoids are identical, it suffices to show that the Lie groupoids $\mathcal{S}_{vac} \rightrightarrows X$, $\mathcal{S}_{vac} \rightrightarrows X^*$ are respectively Lie subgroupoids of $\mathcal{S} \rightrightarrows X$, $\mathcal{S} \rightrightarrows X^*$. For the horizontal structures, it suffices to observe that $X^* \bowtie X \to X_e^* \bowtie X : (g_-, g_+) \mapsto (g_-, Pr_1(g_-), g_+)$ is an injective immersion. The other case is similar. ∎

We now turn to the description of the symplectic properties of the double Lie groupoid $(\mathcal{S}; X^*, X, \mathfrak{h}^*)$ of Theorem 5.2.10. (Recall that $X = \mathfrak{h}^* \times G \times \mathfrak{h}^*$ is the dynamical groupoid for the constant r-matrix $-R$.)

To begin with, a simple computation (using (5.1.4)) shows that the map

$$\rho : (H \times \mathfrak{h}^*) \times (H \times \mathfrak{h}^*) \times G \times G^* \longrightarrow X^* \bowtie X_e$$
$$(h, p, k, q, g, u) \mapsto ((h, p, \varphi_{g^{-1}}^+(u^{-1})^{-1}), (h, k, p, h^{-1}\varphi_{u^{-1}}^-(g^{-1})^{-1}k, q)) \quad (5.2.12)$$

is an isomorphism of groupoids; here the domain is the product groupoid of (5.2.6) and the range is the right action groupoid of Corollary 5.2.8.

Recall that the Poisson bracket of the symplectic groupoid $\mathcal{S}' = (H \times \mathfrak{h}^*) \times (H \times \mathfrak{h}^*)^- \times (G \times G^*) \rightrightarrows X^*$ of (5.2.6) is explicitely given by

$$\{F, F'\}_{\mathcal{S}'}(h, p, k, q, g, u)$$
$$= - <D_1'F', \delta_1 F> + <D_1'F, \delta_1 F'> - <p, [\delta_1 F, \delta_1 F']>$$
$$+ <D_2'F', \delta_2 F> - <D_2'F, \delta_2 F'> + <q, [\delta_2 F, \delta_2 F']> \quad (5.2.13)$$
$$- <\partial F, \lambda_-(DF')(g)> + <\partial_* F, \lambda_+(D_*'F')(u)>$$
$$- <D'F', D_*F> + <D'F, D_*F'>,$$

where the indices 1 and 2 indicate partial derivatives and left/right gradients with respect to the appropriate factor in the first and second copies of $H \times \mathfrak{h}^*$ and the index $*$ indicates partial derivative with respect to G^*. Using the bijection ρ we may transport this Poisson bracket to $\mathcal{S}$ by setting

$$\{F_1, F_2\}_\mathcal{S} \circ \rho = \{F_1 \circ \rho, F_2 \circ \rho\}_{\mathcal{S}'}.$$

Since ρ is a Lie groupoid isomorphism, $(X^* \bowtie X_e, \{,\}_\mathcal{S})$ is a symplectic groupoid.

We now come to the main result of this section.

Theorem 5.2.13. *The double Lie groupoid $(\mathcal{S}; X^*, X, \mathfrak{h}^*)$ where $\mathcal{S}$ is equipped with the Poisson bracket $\{,\}_\mathcal{S}$ is a symplectic double groupoid.*

In order to prove the theorem, it remains to show that $(X_e^* \bowtie X, \{,\}_\mathcal{S})$ is a symplectic groupoid. For this purpose, we will use the isomorphic image $\mathcal{S}' \rightrightarrows X$ of $X_e^* \bowtie X \rightrightarrows X$ under the map ρ and the bracket $\{,\}_{\mathcal{S}'}$. By direct computation, $\mathcal{S}' \rightrightarrows X$ has target and source maps

$$\alpha(h, p, k, q, g, u)$$
$$= (Ad_{h^{-1}}^* p + I^*(\varphi_{g^{-1}}^+(u^{-1})^{-1}), g, Ad_{k^{-1}}^* q + I^*(u)), \quad (5.2.14.a)$$
$$\beta(h, p, k, q, g, u)$$
$$= (p, h^{-1}\varphi_{u^{-1}}^-(g^{-1})^{-1}k, q),$$

5.2. CONSTRUCTION OF THE ASSOCIATED SYMPLECTIC DOUBLE GROUPOID

multiplication map

$$m((h_1,p_1,k_1,q_1,g_1,u_1),(h_2,p_2,k_2,q_2,g_2,u_2)) \\ = (h_1h_2,p_2,k_1k_2,q_2,g_1,u_1\varphi^+_{k_1^{-1}}(u_2)), \tag{5.2.14.b}$$

where

$$(p_1, h_1^{-1}\varphi^-_{u_1^{-1}}(g_1^{-1})^{-1}k_1, q_1) = (Ad^*_{h_2^{-1}}p_2 + I^*(\varphi^+_{g_2^{-1}}(u_2^{-1})^{-1}), g_2, Ad^*_{k_2^{-1}}q_2 + I^*(u_2)),$$

and unit section

$$\epsilon(p,g,q) = (1,p,1,q,g,1). \tag{5.2.14.c}$$

We will verify the conditions of the following proposition of Libermann in [**L**], and show that the unique Poisson structure induced on the base X indeed coincides with that of Theorem 2.1.4.

Proposition 5.2.14. *Let $\Gamma \rightrightarrows P$ be a Lie groupoid equipped with a symplectic form Ω. If Γ is β-connected, the α-foliations and the β-foliations are symplectically orthogonal, and $\epsilon(P) \subset \Gamma$ is Lagrangian, then Γ is a symplectic groupoid over P.*

In our case, that the β-fibers of $\mathcal{S}' \rightrightarrows X$ are connected and $\epsilon(X)$ is Lagrangian are easy to check and we will leave the details to the reader. In order to establish the other condition, we begin with two Propositions which allow us to identify the target and source maps of (5.2.14.a) with canonical projections of natural group actions.

Let $H \times (H \ltimes G^*)$ be the product of H with the Lie group $H \ltimes G^*$ of Proposition 5.1.3 (b)

Proposition 5.2.15. *The left action of $\mathcal{S}' \rightrightarrows X$ on itself induces a left action of the group $H \times (H \ltimes G^*)$ on $\mathcal{S}'$ given by*

$$(h',k',u') \cdot (h,p,k,q,g,u) \\ = (h'h, p, k'k, q, h'\varphi^-_{\varphi^+_{k'}(u')}(g^{-1})^{-1}k'^{-1}, u'\varphi^+_{k'^{-1}}(u)).$$

Moreover, the canonical projection $\mathcal{S}' \longrightarrow H \times (H \ltimes G^*)\backslash \mathcal{S}' \simeq X$ coincides with β.

Proof. Clearly,

$$(h',k',u') \cdot (h,p,k,q,g,u) \\ = m((h',p',k',q',g',u'),(h,p,k,q,g,u))$$

for unique $(p',q',g') \in \mathfrak{h}^* \times \mathfrak{h}^* \times G$ and it is easy to show that this defines a left $H \times (H \ltimes G^*)$ action on $\mathcal{S}'$. Now, if we identify each $H \times (H \ltimes G^*)$-orbit with its unique intersection with $\{1\} \times \mathfrak{h}^* \times \{1\} \times \mathfrak{h}^* \times G \times \{1\}$, then $H \times (H \ltimes G^*)\backslash \mathcal{S}' \simeq X$ and an easy calculation shows the projection map coincides with β. ∎

In a similar way, we have

Proposition 5.2.16. *The right action of $\mathcal{S}' \rightrightarrows X$ on itself induces a right action of the group $H \times (H \ltimes G^*)$ on $\mathcal{S}'$ given by*

$$(h, p, k, q, g, u) \cdot (h', k', u')$$
$$= (hh', Ad_{h'}^* p + I^*(\varphi^+_{k^{-1}\varphi^-_{u^{-1}}(g^{-1})hh'}(u'^{-1})), kk',$$
$$Ad_{k'}^* q + I^*(\varphi^+_{k'}(u'^{-1})), g, u\varphi^+_{k^{-1}}(u')).$$

Moreover, the canonical projection $\mathcal{S}' \longrightarrow \mathcal{S}'/H \times (H \ltimes G^*) \simeq X$ coincides with α. ∎

If $\phi \in C^\infty(X)$, then it follows frow the above propositions that $\alpha^*\phi$ is right $H \times (H \ltimes G^*)$-invariant and $\beta^*\phi$ is left $H \times (H \ltimes G^*)$-invariant. Conversely, it is clear that the right and left $H \times (H \ltimes G^*)$-invariant functions on $\mathcal{S}'$ are of the above form.

In order to show that the α-foliation and the β-foliation are symplectically orthogonal, the next two lemmas are essential.

Lemma 5.2.17. *For $\phi \in C^\infty(X)$, the following formulas hold:*

(a.1) $D'_1(\alpha^*\phi) = ad^*_{\delta_1(\alpha^*\phi)} p$

(a.2) $D'_2(\alpha^*\phi) = ad^*_{\delta_2(\alpha^*\phi)} q$

(a.3) $D'_*(\alpha^*\phi) = Ad_{\phi^-_{u^{-1}}(g^{-1})h} \delta_1(\alpha^*\phi) + Ad_k \delta_2(\alpha^*\phi),$

(b.1) $D_1(\beta^*\phi) + \iota^* D(\beta^*\phi) = 0$

(b.2) $D_2(\beta^*\phi) - \iota^* D'(\beta^*\phi) + \iota^* T_u l_{u^{-1}} \lambda^+ \big(D'_*(\beta^*\phi)\big)(u) = 0$

(b.3) $D_*(\beta^*\phi) - T_g l_{g^{-1}} \lambda^- \big(D(\beta^*\phi)\big)(g) = 0$

Here all partial derivatives as well as left and right gradients are evaluated at $(h, p, k, q, g, u) \in \mathcal{S}'$.

Proof. The formulas in this lemma are the infinitesimal versions of the invariance properties of $\alpha^*\phi$ and $\beta^*\phi$ which can be obtained by using the following basic formulas: $\frac{d}{dt}\big|_{t=0} \varphi^+_{e^{tX}}(u) = \lambda^+(X)(u)$, $\frac{d}{dt}\big|_{t=0} \varphi^-_{e^{t\alpha}}(g) = \lambda^-(\alpha)(g)$, $\frac{d}{dt}\big|_{t=0} \varphi^+_g(e^{t\alpha}) = Ad_g^*\alpha$, $\frac{d}{dt}\big|_{t=0} \varphi^-_u(e^{tX}) = Ad_{u^{-1}}^* X$.

We will illustrate the proof by verifying the identities in (b.1)- (b.3). To do so, we set $h' = e^{tZ_1}, k' = e^{tZ_2}$, and $u' = e^{t\gamma}$ in the relation

$$\beta^*\phi(h'h, p, k'k, q, h'\varphi^-_{\varphi^+_{k'}(u')}(g^{-1})^{-1}k'^{-1}, u\varphi^+_{h'^{-1}}(u))$$
$$= \beta^*\phi(h, p, k, q, g, u),$$

and differentiate with respect to t at $t = 0$. This yields

$$0 = <D_1(\beta^*\phi) + \iota^* D(\beta^*\phi), Z_1>$$
$$+ <D_2(\beta^*\phi) - \iota^* D'(\beta^*\phi), Z_2>$$
$$+ <d_G(\beta^*\phi), -T(l_g \circ r_g)\frac{d}{dt}\big|_{t=0} \varphi^-_{\varphi^+_{e^{tZ_2}}(e^{t\gamma})}(g^{-1})>$$
$$+ <D_*(\beta^*\phi), \gamma> + <d_{G^*}(\beta^*\phi), \frac{d}{dt}\big|_{t=0} \varphi^+_{e^{-tZ_2}}(u)>.$$

Upon using
$$\frac{d}{dt}\Big|_{t=0} \varphi^-_{\varphi^+_{e^{tZ_2}}(e^{t\gamma})}(g^{-1}) = \lambda^-(\gamma)(g^{-1})$$
and the relation
$$-T(l_g \circ r_g)\lambda^-(\gamma)(g^{-1}) = \lambda^-(Ad^*_{g^{-1}}\gamma)(g),$$
this becomes
$$\begin{aligned}0 =& <D_1(\beta^*\phi) + \iota^*D(\beta^*\phi), Z_1> \\ &+ <D_2(\beta^*\phi) - \iota^*D'(\beta^*\phi), Z_2> - <d_{G^*}(\beta^*\phi), \lambda^+(Z_2)(u)> \\ &+ <d_G(\beta^*\phi), \lambda^-(Ad^*_{g^{-1}}\gamma)(g)> + <D_*(\beta^*\phi), \gamma>.\end{aligned}$$

But the two terms above involving $\lambda^\pm$ can be rewritten using the identities
$$<\omega, \lambda^+(Z_2)(u)> = -<\iota^*T_u l_{u^{-1}}\lambda^+(T_1^* l_u \omega)(u), Z_2>$$
and
$$<v, \lambda^-(\gamma)(g)> = -<T_g r_{g^{-1}}\lambda^-(T_1^* r_g v)(g), \gamma>.$$
As $Z_1, Z_2 \in \mathfrak{h}$ and $\gamma \in \mathfrak{g}^*$ are independent, we obtain (b.1)- (b.3). ∎

Lemma 5.2.18. (a) For all $g \in G, u \in G^*$ and $Z \in \mathfrak{h}$,
$$Ad^*_{u^{-1}}Ad_{\varphi^-_{u^{-1}}(g^{-1})}Z = Ad_{g^{-1}}Z + T_g l_{g^{-1}}\lambda^-\big(Tr_{\varphi^+_{g^{-1}}(u^{-1})}\lambda^+(Z)(\varphi^+_{\varphi^-_{u^{-1}}(g^{-1})}(u))\big)(g).$$

(b) $\varphi^+_{g^{-1}} \circ r_{u^{-1}} \circ \varphi^+_{\varphi^-_{u^{-1}}(g^{-1})^{-1}} = r_{\varphi^+_{g^{-1}}(u^{-1})}$
for all $g \in G, u \in G^*$.

Proof.

(a) We have
$$Ad^*_{u^{-1}} Ad_{\varphi^-_{u^{-1}}(g^{-1})}Z$$
$$= \frac{d}{dt}\Big|_{t=0} \varphi^-_u\big(exp(tAd_{\varphi^-_{u^{-1}}(g^{-1})}Z)\big).$$

Then, by repeated application of (5.1.4) and the triviality of the action of G^* on H, the desired identity follows. We will skip the intermediate steps as the notation gets clumsy.

(b)
$$\begin{aligned}&\varphi^+_{g^{-1}} \circ r_{u^{-1}} \circ \varphi^+_{\varphi^-_{u^{-1}}(g^{-1})^{-1}}(u') \\ &= \varphi^+_{g^{-1}}(\varphi^+_{\varphi^-_{u^{-1}}(g^{-1})^{-1}}(u')\, u^{-1}) \\ &= \varphi^+_{\varphi^-_{u^{-1}}(g^{-1})}(\varphi^+_{\varphi^-_{u^{-1}}(g^{-1})^{-1}}(u'))\varphi^+_{g^{-1}}(u^{-1})\end{aligned}$$
by (5.1.4)
$$= u'\varphi^+_{g^{-1}}(u^{-1}).$$

Hence the assertion. ∎

Proposition 5.2.19. (Polarity condition.) For all $\phi, \psi \in C^\infty(X)$,
$$\{\alpha^*\phi, \beta^*\psi\}_{\mathcal{S'}} = 0.$$

54   5. THE CONSTANT R-MATRIX CASE

Proof. Let $\hat{\phi} = \alpha^*\phi$, $\hat{\psi} = \beta^*\psi$. By invoking the identities (a.1), (a.2) and (b.3) of Lemma 5.2.17, we have

$$\{\hat{\phi}, \hat{\psi}\}_{\mathcal{S}'}(h, p, k, q, g, u)$$
$$= - <D_1\hat{\psi}, Ad_h\delta_1\hat{\phi}> + <D_2\hat{\psi}, Ad_k\delta_2\hat{\phi}>$$
$$+ <d_*\hat{\phi}, \lambda^+(D'_*\hat{\psi})(u)> - <D'\hat{\psi}, D_*\hat{\phi}>.$$

Next, using (b.1), (b.2) and (b.3) of the same lemma successively gives

$$\{\hat{\phi}, \hat{\psi}\}_{\mathcal{S}'}(h, p, k, q, gu)$$
$$= <D'\hat{\psi}, Ad_{g^{-1}h}\delta_1\hat{\phi} + Ad_k\delta_2\hat{\phi} - D_*\hat{\phi}$$
$$+ T_gl_{g^{-1}}\lambda^-(Ad^*_{g^{-1}}T_ur_{u^{-1}}\lambda^+(Ad_{\varphi^-_{u^{-1}}(g^{-1}h)}\delta_1\hat{\phi})(u))(g)>$$

Now using (a.3) together with $Ad^*_{u^{-1}}Z = Z$, $\forall Z \in \mathfrak{h}$, we obtain

$$\{\hat{\phi}, \hat{\psi}\}_{\mathcal{S}'}(h, p, k, q, g, u)$$
$$= <D'\hat{\psi}, Ad_{g^{-1}h}\delta_1\hat{\phi} - Ad^*_{u^{-1}}Ad_{\varphi^-_{u^{-1}}(g^{-1})h}\delta_1\hat{\phi}$$
$$+ T_gl_{g^{-1}}\lambda^-(Ad^*_{g^{-1}}T_ur_{u^{-1}}\lambda^+(Ad_{\varphi^-_{u^{-1}}(g^{-1})h}\delta_1\hat{\phi})(u))(g)>$$
$$= <D'\hat{\psi}, T_gl_{g^{-1}}\lambda^-(Ad^*_{g^{-1}}T_ur_{u^{-1}}\lambda^+(Ad_{\varphi^-_{u^{-1}}(g^{-1})h}\delta_1\hat{\phi})(u))$$
$$- Tr_{\varphi^+_{g^{-1}}(u^{-1})}\lambda^+(Ad_h\delta_1\hat{\phi})(\varphi^+_{\varphi^-_{u^{-1}}(g^{-1})}(u)))(g)>$$

where in the last equality we have used Lemma 5.2.18 (a).
But

$$\lambda_+(Ad_{\varphi^-_{u^{-1}}(g^{-1})h}\delta_1\hat{\phi})(u)$$
$$= T\varphi^+_{\varphi^-_{u^{-1}}(g^{-1})^{-1}}\lambda_+(Ad_h\delta_1\hat{\phi})(\varphi^+_{g^{-1}}(u^{-1})^{-1}),$$

hence the assertion that $\{\hat{\phi}, \hat{\psi}\}_{\mathcal{S}'} = 0$ now follows from Lemma 5.18 (b). ∎

To prepare for the proof that α is a Poisson map, we have to compute the explicit expressions for the partial derivatives of $\alpha^*\phi$, $\phi \in C^\infty(X)$, as well as its right gradients. We also establish further identities for the left dressing vector field.

Lemma 5.2.20. *For $\phi \in C^\infty(X)$,*

(a) $\delta_1(\alpha^*\phi) = Ad_{h^{-1}}\delta_1\phi(\alpha(s))$,

(b) $\delta_2(\alpha^*\phi) = Ad_{k^{-1}}\delta_2\phi(\alpha(s))$,

(c) $D(\alpha^*\phi) = D\phi(\alpha(s)) - Tl_{\varphi^+_{g^{-1}}(u^{-1})^{-1}}\lambda^+(\delta_1\phi(\alpha(s)))(\varphi^+_{g^{-1}}(u^{-1}))$,

(d) $D_*(\alpha^*\phi) = \delta_2\phi(\alpha(s)) + Ad^*_{u^{-1}}Ad_{\varphi^-_{u^{-1}}(g^{-1})}\delta_1\phi(\alpha(s))$

$\qquad = \delta_2\phi(\alpha(s)) + Ad_{g^{-1}}\delta_1\phi(\alpha(s))$

$\qquad -T_gl_{g^{-1}}\lambda^-\bigl(Tl_{\varphi^+_{g^{-1}}(u^{-1})^{-1}}\lambda^+(\delta_1\phi(\alpha(s)))(\varphi^+_{g^{-1}}(u^{-1}))\bigr)(g)$ (*second form*).

Here, all the partial derivatives and right gradients of $\alpha^*\phi$ are evaluated at $s = (h, p, k, q, g, u) \in \mathcal{S}'$.

5.2. CONSTRUCTION OF THE ASSOCIATED SYMPLECTIC DOUBLE GROUPOID

Proof. We will establish the formulas in (d), leaving the other parts to the interested reader. For $\gamma \in \mathfrak{g}^*$, we have

$$< D'_*(\alpha^*\phi), \gamma >$$
$$= \frac{d}{dt}\Big|_{t=0} \phi\big(Ad^*_{h^{-1}}p - I^*(\varphi^+_{g^{-1}}(e^{-t\gamma}u^{-1})), g, Ad^*_{k^{-1}}q + I^*(ue^{t\gamma})\big)$$
$$= < \delta_1\phi(\alpha(s)), -TI^* \frac{d}{dt}\Big|_0 \varphi^+_{g^{-1}}(e^{-t\gamma}u^{-1}) > + < \delta_2\phi, \iota^*\gamma > .$$

Since
$$\frac{d}{dt}\Big|_{t=0} \varphi^+_{g^{-1}}(e^{-t\gamma}u^{-1}) = -T_1 r^+_{\varphi^+_{g^{-1}}(u^{-1})} Ad^*_{\varphi^-_{u^{-1}}(g^{-1})}\gamma,$$

and
$$T_1(I^* \circ r^+_{\varphi^+_{g^{-1}}(u^{-1})}) = T_1 I^* = \iota^*,$$

it follows that
$$D'_*(\alpha^*\phi) = \delta_2\phi(\alpha(s)) + Ad_{\varphi^-_{u^{-1}}(g^{-1})}\delta_1\phi(\alpha(s)),$$

and this gives the first form of $D_*(\alpha^*\phi)$. To obtain the second form of $D_*(\alpha^*\phi)$ from the first one, simply apply Lemma 5.2.18 (a) and the fact that

$$\lambda^+(Z)(\varphi^+_{g^{-1}}(u^{-1})^{-1})$$
$$= -T\big(l_{\varphi^+_{g^{-1}}(u^{-1})^{-1}} \circ r_{\varphi^+_{g^{-1}}(u^{-1})^{-1}}\big)\lambda^+(Z)(\varphi^+_{g^{-1}}(u^{-1})), \quad Z \in \mathfrak{h}. \blacksquare$$

Lemma 5.2.21. *For all $g \in G$, $u \in G^*$, and $Z \in \mathfrak{h}$,*

(a) $T_u I^* \lambda^+(Z)(u) = ad^*_Z I^*(u),$

(b) $Ad^*_g Tl_{\varphi^+_{g^{-1}}(u^{-1})^{-1}} \lambda^+(Z)(\varphi^+_{g^{-1}}(u^{-1}))$
$= -T_u r_{u^{-1}} \lambda^+(Ad_{\varphi^-_{u^{-1}}(g^{-1})}Z)(u),$

(c) $Tl_{\varphi^+_{g^{-1}}(u^{-1})^{-1}} \lambda^+(Ad_g Z)(\varphi^+_{g^{-1}}(u^{-1}))$
$= -Ad_{\varphi^+_{g^{-1}}(u^{-1})^{-1}} Ad^*_{\varphi^-_{u^{-1}}(g^{-1})} T_u l_{u^{-1}} \lambda^+(Z)(u).$

Proof. (a) This is the infinitesimal version of the $H-$ equivariance of the map I^*.

(b)
$$\lambda^+(Ad_{\varphi^-_{u^{-1}}(g^{-1})}Z)(u)$$
$$= \frac{d}{dt}\Big|_{t=0} \varphi^+_{exp\big(tAd_{\varphi^-_{u^{-1}}(g^{-1})}Z\big)}(u)$$
$$= T\varphi^+_{\varphi^-_{u^{-1}}(g^{-1})^{-1}} \lambda^+(Z)(\varphi^+_{g^{-1}}(u^{-1})^{-1})$$
$$= -T\varphi^+_{\varphi^-_{u^{-1}}(g^{-1})^{-1}} Tr_{\varphi^+_{g^{-1}}(u^{-1})^{-1}} Tl_{\varphi^+_{g^{-1}}(u^{-1})^{-1}} \lambda^+(Z)(\varphi^+_{g^{-1}}(u^{-1}))$$

(where we have used $Ad^*_u Z = Z$ for $u \in G^*, Z \in \mathfrak{h}$).

Now $Ad_g^* = T_1\varphi_g^+$ and by a straightforward calculation using (5.1.4), we find

$$\varphi_g^+ \circ r_{\varphi_{g^{-1}}^+(u^{-1})}^+ \circ \varphi_{\varphi_{u^{-1}}^-(g^{-1})}^+ = r_{u^{-1}}.$$

Hence the assertion follows.

(c) We have

$$\lambda^+(Ad_g Z)(\varphi_{g^{-1}}^+(u^{-1}))$$
$$= \frac{d}{dt}\bigg|_{t=0} \varphi_{e^{tAd_g Z}}^+(\varphi_{g^{-1}}^+(u^{-1}))$$
$$= \frac{d}{dt}\bigg|_{t=0} \varphi_{g^{-1}}^+\left(\left(\varphi_{\varphi_{u^{-1}}^-(e^{tZ})}^+(u)\right)^{-1}\right)$$
$$= \frac{d}{dt}\bigg|_{t=0} \varphi_{g^{-1}}^+(\varphi_{e^{tZ}}^+(u)^{-1})$$

by the triviality of the action of G^* on H

$$= -T_{u^{-1}}\varphi_{g^{-1}}^+ T_1 r_{u^{-1}} T_u l_{u^{-1}} \lambda^+(Z)(u).$$

On the other hand, for $\tilde{u} \in G^*$, we find

$$l_{\varphi_{g^{-1}}^+(u^{-1})^{-1}} \circ \varphi_{g^{-1}}^+ \circ r_{u^{-1}}(\tilde{u}) = \varphi_{g^{-1}}^+(u^{-1})^{-1} \varphi_{\varphi_{u^{-1}}^-(g^{-1})}^+(\tilde{u}) \varphi_{g^{-1}}^+(u^{-1})$$

from which we deduce that

$$T_1\left(l_{\varphi_{g^{-1}}^+(u^{-1})^{-1}} \circ \varphi_{g^{-1}}^+ \circ r_{u^{-1}}\right) = Ad_{\varphi_{g^{-1}}^+(u^{-1})^{-1}} Ad^*_{\varphi_{u^{-1}}^-(g^{-1})}.$$

The assertion is now clear. ■

Proposition 5.2.22. *The map* $\alpha : (S', \{\,,\,\}_{S'}) \longrightarrow (X, \{\,,\,\}_X)$ *is a Poisson map.*

Proof. We want to show

$$\{\alpha^*\phi, \alpha^*\psi\}_{S'} = \alpha^*\{\phi, \psi\}_X$$

for all $\phi, \psi \in C^\infty(X)$. We will evaluate $\{\alpha^*\phi, \alpha^*\psi\}_{S'}$ at $s = (h, p, k, q, g, u) \in S'$ and set $x = \alpha(s)$.

By using Lemma 5.2.17 (a.1), (a.2) and Lemma 5.2.20 (a), (b), we have

$$\{\alpha^*\phi, \alpha^*\psi\}_{S'}(s)$$
$$= <Ad_{h^{-1}}^* p, [\delta_1\phi(x), \delta_1\psi(x)]> - <Ad_{k^{-1}}^* q, [\delta_2\phi(x), \delta_2\psi(x)]>$$
$$- <D'(\alpha^*\phi), T_g l_{g^{-1}} \lambda^-(D(\alpha^*\psi))(g) - D_*(\alpha^*\psi)>$$
$$+ <D_*(\alpha^*\phi), T_u r_{u^{-1}} \lambda^+(D'_*(\alpha^*\psi))(u) - D'(\alpha^*\psi)>$$

Now, it follows from Lemma 5.2.20 (c) and (d) (second form) that

$$T_g l_{g^{-1}} \lambda^-(D(\alpha^*\psi))(g) - D_*(\alpha^*\psi)$$
$$= T_g l_{g^{-1}} \lambda^-(D\psi(x))(g) - \delta_2\psi(x) - Ad_{g^{-1}}\delta_1\psi(x).$$

5.2. CONSTRUCTION OF THE ASSOCIATED SYMPLECTIC DOUBLE GROUPOID

On the other hand, by using Lemma 5.2.20 (c), (d) (first form) and Lemma 5.2.21 (b), we obtain

$$T_u r_{u^{-1}} \lambda^+(D'_*(\alpha^*\psi))(u) - D'(\alpha^*\psi)$$
$$= T_u r_{u^{-1}} \lambda^+(\delta_2 \psi(x))(u) - D'\psi(x).$$

Consequently,

$$- < D'(\alpha^*\phi), T_g l_{g^{-1}} \lambda^-(D(\alpha^*\psi))(g) - D_*(\alpha^*\psi) >$$
$$+ < D_*(\alpha^*\phi), T_u r_{u^{-1}} \lambda^+(D'_*(\alpha^*\psi))(u) - D'(\alpha^*\psi) >$$
$$= < D'\phi(x), \delta_2 \psi(x) > + < D\phi(x), \delta_1 \psi(x) >$$
$$- < D'\psi(x), \delta_2 \phi(x) > - < D\psi(x), \delta_1 \phi(x) >$$
$$- < D'\phi(x), T_g l_{g^{-1}} \lambda^-(D\psi(x))(g) >$$
$$+ T_1 + T_2 + T_3 + T_4,$$

where

$$T_1 = < Ad_g^* Tl_{\varphi^+_{g^{-1}}(u^{-1})^{-1}} \lambda^+(\delta_1\phi(x))(\varphi^+_{g^{-1}}(u^{-1})), T_g l_{g^{-1}} \lambda^-(D\psi(x))(g) >$$
$$+ < T_g l_{g^{-1}} \lambda^-(Tl_{\varphi^+_{g^{-1}}(u^{-1})^{-1}} \lambda^+(\delta_1\phi(x))(\varphi^+_{g^{-1}}(u^{-1})))(g), D'\psi(x) >,$$

$$T_2 = - < \iota^* Tl_{\varphi^+_{g^{-1}}(u^{-1})^{-1}} \lambda^+(\delta_1\phi(x))(\varphi^+_{g^{-1}}(u^{-1})), \delta_1\psi(x) >,$$

$$T_3 = < \delta_1\phi(x), \iota^* T_u r_{u^{-1}} \lambda^+(\delta_2\psi(x))(u) >,$$
$$T_4 = < Ad_{u^{-1}}^* Ad_{\varphi^-_{u^{-1}}(g^{-1})} \delta_1\phi(x), T_u r_{u^{-1}} \lambda^+(\delta_2\psi(x))(u) >$$
$$- < Ad_g^* Tl_{\varphi^+_{g^{-1}}(u^{-1})^{-1}} \lambda^+(\delta_1\phi(x))(\varphi^+_{g^{-1}}(u^{-1})), \delta_2\psi(x) > .$$

Using the relation

$$< \lambda^-(\gamma)(g), v > = - < \gamma, T_g r_{g^{-1}} \lambda^-(T_1^* r_g v)(g) >,$$

it is immediate that $T_1 = 0$. For the term T_2, note that

$$\iota^* Tl_{\varphi^+_{g^{-1}}(u^{-1})^{-1}} = T_{\varphi^+_{g^{-1}}(u^{-1})} I^*.$$

Hence it follows from Lemma 5.2.21 (a) that

$$T_2 = - < ad^*_{\delta_1\phi(x)} I^*(\varphi^+_{g^{-1}}(u^{-1})), \delta_1\psi(x) >$$
$$= - < I^*(\varphi^+_{g^{-1}}(u^{-1})), [\delta_1\phi(x), \delta_1\psi(x)] > .$$

Similarly we have

$$T_3 = - < I^*(u), [\delta_2\phi(x), \delta_2\psi(x)] > .$$

Assembling the calculations, we find

$$\{\alpha^*\phi, \alpha^*\psi\}_{S'}(s) = \{\phi, \psi\}_X(x) + T_4.$$

Hence it remains to show that $T_4 = 0$. To do so, we invoke the relation

$$< \lambda^+(X)(u), \omega > = - < X, T_u l_{u^{-1}} \lambda^+(T_1^* l_u \omega)(u) >$$

and Lemma 5.2.21 (c) to rewrite T_4 as

$$T_4 = <\delta_1\phi(x), -\iota^* Ad_{\varphi^+_{g^{-1}}(u^{-1})^{-1}} Ad^*_{\varphi^-_{u^{-1}}(g^{-1})} T_u l_{u^{-1}} \lambda^+(\delta_2\psi(x))(u)$$
$$+ \iota^* Ad^*_{\varphi^-_{u^{-1}}(g^{-1})} T_u l_{u^{-1}} \lambda^+(\delta_2\psi(x))(u) >.$$

But from the triviality of the $Ad^*_{G^*}$ action on $\mathfrak{h}$, it follows that

$$\iota^* Ad^*_{\varphi^-_{u^{-1}}(g^{-1})} = \iota^* Ad^*_{\varphi^+_{g^{-1}}(u^{-1})^{-1}} Ad^*_{\varphi^-_{u^{-1}}(g^{-1})}.$$

Therefore, $T_4 = 0$. ∎

Combining the above Proposition with Proposition 5.2.16, we have

Corollary 5.2.23. *The right action of $H \times (H \ltimes G^*)$ on $\mathcal{S}'$ in Proposition 5.2.16 is admissible, i.e. functions in $C^\infty(\mathcal{S}')$ invariant under the action form a Lie subalgebra of $C^\infty(\mathcal{S}')$. Furthermore, the quotient Poisson structure on $X \simeq \mathcal{S}'/H \times (H \ltimes G^*)$ coincides with $\{\,,\,\}_X$.* ∎

We will skip the proof of the next two lemmas. They play an important role in showing that β is an anti-Poisson map.

Lemma 5.2.24. *For $\phi \in C^\infty(X)$, the following formulas hold:*

(a) $\delta_1(\beta^*\phi) = \delta_1\phi(\beta(s))$,

(b) $\delta_2(\beta^*\phi) = \delta_2\phi(\beta(s))$,

(c) $D'_1(\beta^*\phi) = -\iota^* D\phi(\beta(s))$,

(d) $D'_2(\beta^*\phi) = \iota^* D'\phi(\beta(s))$,

(e) $D(\beta^*\phi) = Ad_{\varphi^+_{g^{-1}}(u^{-1})^{-1}} Ad^*_{h^{-1}} D\phi(\beta(s))$,

(f) $D'_*(\beta^*\phi) = -Tr_{\varphi^-_{u^{-1}}(g^{-1})^{-1}} \lambda^-(Ad^*_{k^{-1}} D'\phi(\beta(s)))(\varphi^-_{u^{-1}}(g^{-1}))$.

Here, all the partial derivatives as well as left and right gradients of $\beta^*\phi$ are evaluated at $s = (h, p, k, q, g, u) \in \mathcal{S}'$. ∎

Lemma 5.2.25. *For all $g \in G, u \in G^*$, and $\gamma \in \mathfrak{g}^*$,*

$$Ad^*_g Ad^*_{\varphi^+_{g^{-1}}(u^{-1})^{-1}} \gamma = Ad_u Ad^*_{\varphi^-_{u^{-1}}(g^{-1})^{-1}} \gamma$$
$$- T_u r_{u^{-1}} \lambda^+(Tr_{\varphi^-_{u^{-1}}(g^{-1})^{-1}} \lambda^-(Ad^*_{\varphi^-_{u^{-1}}(g^{-1})^{-1}} \gamma)(\varphi^-_{u^{-1}}(g^{-1})))(u). ∎$$

Proposition 5.2.26. *The map $\beta : (\mathcal{S}', \{\,,\,\}_{\mathcal{S}'}) \longrightarrow (X, \{\,,\,\}_X)$ is an anti-Poisson map.*

Proof. Let $\phi, \psi \in C^\infty(X)$, $s = (h, p, k, q, g, u) \in \mathcal{S}'$ and denote $\beta(s)$ by x. From Lemma 5.2.24 (a) - (d) and Lemma 5.2.17 (b.1), we immediately have

$$\{\beta^*\phi, \beta^*\psi\}_{\mathcal{S}'}(s)$$
$$= <\iota^* D\psi(x), \delta_1\phi(x)> - <\iota^* D\phi(x), \delta_1\psi(x)> - <p, [\delta_1\phi(x), \delta_1\psi(x)]>$$
$$+ <\iota^* D'\psi(x), \delta_2\phi(x)> - <\iota^* D'\phi(x), \delta_2\psi(x)> + <q, [\delta_2\phi(x), \delta_2\psi(x)]>$$
$$+ <D'_*(\beta^*\phi), T_u l_{u^{-1}} \lambda^+(D'_*(\beta^*\psi))(u) - Ad_{u^{-1}} D'(\beta^*\psi)>.$$

5.2. CONSTRUCTION OF THE ASSOCIATED SYMPLECTIC DOUBLE GROUPOID   59

Hence we have to show that

$$< D'_*(\beta^*\phi), T_u l_{u^{-1}} \lambda^+(D'_*(\beta^*\psi))(u) - Ad_{u^{-1}} D'(\beta^*\psi) >$$
$$=< d_G\phi(x), \Pi_G^\#(h^{-1}\varphi^-_{u^{-1}}(g^{-1})^{-1}k) d_G\psi(x) >.$$

To do so, we apply Lemma 5.2.24 (e), (f) and Lemma 5.2.25, this yields

$$T_u l_{u^{-1}} \lambda^+(D'_*(\beta^*\psi))(u) - Ad_{u^{-1}} D'(\beta^*\psi)$$
$$= -Ad^*_{\varphi^-_{u^{-1}}(g^{-1})^{-1}} Ad^*_{h^{-1}} D\psi(x).$$

Consequently,

$$< D'_*(\beta^*\phi), T_u l_{u^{-1}} \lambda^+(D'_*(\beta^*\psi))(u) - Ad_{u^{-1}} D'(\beta^*\psi) >$$
$$=< T r_{\varphi^-_{u^{-1}}(g^{-1})} \lambda^-(Ad^*_{k^{-1}} D'\phi(x))(\varphi^-_{u^{-1}}(g^{-1})), Ad^*_{\varphi^-_{u^{-1}}(g^{-1})^{-1}} Ad^*_{h^{-1}} D\psi(x) >$$
$$=< d_G\psi(x), T_{\varphi^-_{u^{-1}}(g^{-1})}(l_{h^{-1}\varphi^-_{u^{-1}}(g^{-1})^{-1}} \circ r_{\varphi^-_{u^{-1}}(g^{-1})^{-1}k}) \Pi_G^\#(\varphi^-_{u^{-1}}(g^{-1}))$$
$$T^*_{\varphi^-_{u^{-1}}(g^{-1})}(l_{h^{-1}\varphi^-_{u^{-1}}(g^{-1})^{-1}} \circ r_{\varphi^-_{u^{-1}}(g^{-1})^{-1}k}) d_G\phi(x) >$$
$$= -< d_G\psi(x), \Pi_G^\#(h^{-1}\varphi^-_{u^{-1}}(g^{-1})^{-1}k) d_G\phi(x) >,$$

where in the last step, we have used the fact that Π_G is multiplicative and that Π_G vanishes on H. This completes the proof. ∎

As a consequence of the Proposition 5.2.26 and Proposition 5.2.15, we obtain

Corollary 5.2.27. *The left action of $H \times (H \ltimes G^*)$ on $\mathcal{S}'$ in Proposition 5.2.15 is admissible (see Cor.5.2.23). Furthermore, the quotient Poisson structure on $X \simeq H \times (H \ltimes G^*)\backslash \mathcal{S}'$ coincides with $-\{,\}_X$.* ∎

This completes the proof that $\mathcal{S}' \rightrightarrows X$ is a symplectic groupoid.

Remark 5.2.28. For the case where $R = 0$, we can show that the horizontal structure $\mathcal{S}' \rightrightarrows X$ is a symplectic groupoid by using the fact that the conormal bundle $N^*(B)$ of a submanifold $B \subset M$ is a Lagrangian submanifold of T^*M, equipped with its canonical symplectic structure. We will present the simple argument here. For this purpose, we introduce the submanifold

$$\mathcal{C} = \{(h_1, k_1, g_1, h_2, k_2, h_1^{-1}g_1k_1, h_1h_2, k_1k_2, g_1) \mid$$
$$h_1, h_2, k_1, k_2 \in H, g_1 \in G\} \subset (H \times H \times G)^3,$$

and let $\mathcal{M}$ denote the graph of the multiplication of the horizontal structure. We want to show that $\mathcal{M}$ is a Lagrangian submanifold of $((H \times \mathfrak{h}^*) \times (H \times \mathfrak{h}^*)^- \times (G \times \mathfrak{g}^*))^2 \times ((H \times \mathfrak{h}^*) \times (H \times \mathfrak{h}^*)^- \times (G \times \mathfrak{g}^*))^-$. If

$$\tau : (T^*H \times T^*H \times T^*G)^3 \longrightarrow ((H \times \mathfrak{h}^*) \times (H \times \mathfrak{h}^*) \times (G \times \mathfrak{g}^*))^3$$

is the trivialization of $(T^*H \times T^*H \times T^*G)^3$ by left translation, then equivalently, we just have to show $\tau^{-1}(\mathcal{M})$ is a Lagrangian submanifold of $(T^*H \times (T^*H)^- \times T^*G)^2 \times (T^*H \times (T^*H)^- \times T^*G)^-$. Let $m_{-1} : T^*H \longrightarrow (T^*H)^-$, and $\mu_{-1} : T^*H \times (T^*H)^- \times T^*G \longrightarrow (T^*H \times (T^*H)^- \times T^*G)^-$ be multiplication by -1, and denote by $\tilde{Id}$ the identity map on $T^*H \times (T^*H)^- \times T^*G$. Then by a direct calculation, we find that

$$\tau^{-1}(\mathcal{M}) = (\tilde{Id}, \tilde{Id}, \mu_{-1}) \circ (Id_{T^*H}, m_{-1}, Id_{T^*G})^3 N^*(\mathcal{C}),$$

where $N^*(\mathcal{C})$ is the conormal bundle of $\mathcal{C}$. As m_{-1} and μ_{-1} are symplectomorphisms, this establishes the assertion.

We now turn to the description of the symplectic foliation of X.

Equip $H \times (H \ltimes G^*)$ with the product of the trivial Poisson Lie group structure on H and the Poisson Lie group structure of Proposition 5.1.3 on $H \ltimes G^*$. It is easy to see that the groupoid action Ψ^- of (5.2.8) restricts to a group action

$$\tilde{\Psi}^- : H \times (H \ltimes G^*) \times X \longrightarrow X,$$

given by

$$((k,(h,u)),(p,g,q)) \mapsto \Psi^-_{(h,p,u,k)}(p,g,q).$$

Theorem 5.2.29. *(a)* $\tilde{\Psi}^-$ *is a left Poisson Lie group action.*

(b) The symplectic leaf $\mathcal{L}_{(p,g,q)}$ *in* $(X, \{\,,\,\}_X)$ *passing through the point* (p,g,q) *is the orbit of* (p,g,q) *under the action* $\tilde{\Psi}^-$ *i.e.*

$$\mathcal{L}_{(p,g,q)} = \{(Ad^*_{h^{-1}}p + I^*(u), \varphi^-_u(hgk^{-1}), Ad^*_{k^{-1}}q + I^*(\varphi^+_{hgk^{-1}}(u)))$$
$$\mid (k,(h,u)) \in H \times (H \ltimes G^*)\}.$$

Proof. (a) Equip X with the Poisson bracket of Theorem 2.1.4 (with r-matrix $-R$) and $H \times (H \ltimes G^*)$ with the Poisson Lie bracket

$$\{\phi, \psi\}(k,(h,u)) = \partial_*\phi(u)\,\Pi_*(u)\partial_*\psi(u)$$

of Proposition 5.1.3. We have to show that the action

$$\tilde{\Psi}^- : \quad H \times (H \ltimes G^*) \times X \quad \longrightarrow \quad X$$
$$((k,(h,u)),(p,g,q)) \mapsto (Ad^*_{h^{-1}}p + I^*(u), \varphi^-_u(hgk^{-1}), Ad^*_{k^{-1}}q + I^*(\varphi^+_{hgk^{-1}}(u)))$$

satisfies

$$\{f \circ \tilde{\Psi}^-, f' \circ \tilde{\Psi}^-\}_{H \times (H \ltimes G^*) \times X} = \{f, f'\}_X \circ \tilde{\Psi}^-. \qquad (P)$$

To begin with, a direct calculation making use of the equivariance of I^*, (5.1.4), and the triviality of the action of G^* on H yields the following expressions for the partial derivatives of $f \circ \tilde{\Psi}^-$ at $((k,(h,u)),p,g,q) \in H \times (H \ltimes G^*) \times X$:

$$\partial_*(f \circ \tilde{\Psi}^-) = T^*_u r_{u^{-1}}\big(\iota\delta_1 f - \Pi^r(\varphi^-_u(hgk^{-1}))Df + Ad_{\varphi^-_u(hgk^{-1})}\iota\delta_2 f\big),$$
$$\delta_1(f \circ \tilde{\Psi}^-) = Ad_{h^{-1}}\delta_1 f,$$
$$\delta_2(f \circ \tilde{\Psi}^-) = Ad_{k^{-1}}\delta_2 f,$$
$$\partial(f \circ \tilde{\Psi}^-) = T^*_g l_{g^{-1}}\big(Ad_{\varphi^+_{hg}(u)^{-1}}Ad^*_k D'f + \Pi^l_*(\varphi^+_{hg}(u))\iota Ad_{k^{-1}}\delta_2 f\big),$$

where Π^r (resp. Π^l_*) stands for the Poisson tensor of G (resp. G^*) in the right (resp. left) invariant frame.

5.2. CONSTRUCTION OF THE ASSOCIATED SYMPLECTIC DOUBLE GROUPOID

We will restrict ourselves to an outline of the main steps of the calculation of $lhs(P) - rhs(P)$. We use the shorthand notation d_{ij}, $i, j \in \{1, 2, *\}$ to stand for $\{p_i^* \phi \circ \tilde{\Psi}^-, p_j^* \psi \circ \tilde{\Psi}^-\}_{H \times (H \ltimes G^*) \times X} - \{p_i^* \phi, p_j^* \psi\}_X \circ \tilde{\Psi}^-$ where p_i is as in the proof of Propositon 2.2.4.

We now compute d_{ij} for the various cases.

(1) $d_{11} = <\iota\delta\phi, \Pi_*^r(u)\iota\delta\psi> - <I^*(u), [\delta\phi, \delta\psi]> = 0$ by Lemma 5.2.21 (a).

(2)
$$d_{22} = -2 <\iota Ad_{k^{-1}}\delta\phi, \Pi_*^l(\varphi^+_{hg}(u))\iota Ad_{k^{-1}}\delta\psi> + <\iota\delta\phi, \Pi_*^l(\varphi^+_{hgk^{-1}}(u))\iota\delta\psi>$$
$$+ <Ad_{\varphi_u^-(hgk^{-1})}\iota\delta\phi, \Pi_*^r(u)Ad_{\varphi_u^-(hgk^{-1})}\iota\delta\psi>$$
$$+ <\iota Ad_{k^{-1}}\delta\phi, \Pi_*^l(\varphi^+_{hg}(u))\Pi^l(g)\Pi_*^l(\varphi^+_{hg}(u))\iota Ad_{k^{-1}}\delta\psi>.$$

Using the fact that the action $H \times G^* \longrightarrow G^* : (k, v) \mapsto \varphi^+_{k^{-1}}(v)$ is Hamiltonian, we have

$$\Pi_*^l(\varphi^+_{hgk^{-1}}(u)) = \Pi_*^l(\varphi^+_{k^{-1}}(\varphi^+_{hg}(u))) = Ad_{k^{-1}}^* \Pi_*^l(\varphi^+_{hg}(u)) Ad_{k^{-1}} \qquad (E.1)$$

while, since Π is multiplicative and vanishes on $H \subset G$, we obtain

$$\Pi^l(hgk^{-1}) = Ad_k \Pi^l(g) Ad_k^*. \qquad (E.2)$$

Therefore,
$$d_{22} = -<\iota\delta\phi, \Pi_*^l(\varphi^+_{hgk^{-1}}(u))\iota\delta\psi>$$
$$+ <Ad_{\varphi_u^-(hgk^{-1})}\iota\delta\phi, \Pi_*^r(u)Ad_{\varphi_u^-(hgk^{-1})}\iota\delta\psi>$$
$$+ <\iota\delta\phi, \Pi_*^l(\varphi^+_{hgk^{-1}}(u))\Pi^l(hgk^{-1})\Pi_*^l(\varphi^+_{hgk^{-1}}(u))\iota\delta\psi>.$$

Next, observe that the Poisson property of $\varphi^+ : G^* \times \overline{G} \longrightarrow G^*$ may be written as

$$Ad^*_{\varphi_u^-(x)} \Pi_*^r(u) Ad_{\varphi_u^-(x)}$$
$$= Ad^*_{\varphi_x^+(u)}\big(-\Pi_*^l(\varphi_x^+(u))\Pi^l(x)\Pi_*^l(\varphi_x^+(u)) + \Pi_*^l(\varphi_x^+(u))\big)Ad^*_{\varphi_x^+(u)}.$$

Thus, $d_{22} = 0$ follows from $Ad_v^* \iota Z = \iota Z$, $Z \in \mathfrak{h}$, $v \in G^*$.

(3)
$$d_{12} = <\iota\delta\phi, \big(\Pi_*^r(u)Ad_{\varphi_u^-(hgk^{-1})} - Ad^*_{g^{-1}h^{-1}}\Pi_*^l(\varphi^+_{hg}(u))Ad_{k^{-1}}\big)\iota\delta\psi>$$
$$= <\iota\delta\phi, \big(\Pi_*^r(u)Ad_{\varphi_u^-(hgk^{-1})} - Ad^*_{(hgk^{-1})^{-1}}\Pi_*^l(\varphi^+_{hgk^{-1}}(u))\big)\iota\delta\psi>.$$

But it follows from Lemma 5.2.21 (b) that

$$Ad^*_{x^{-1}}\Pi_*^l(\varphi_x^+(u))\iota Z = Ad_{u^{-1}}\Pi_*^r(u)Ad_{\varphi_u^-(x)}\iota Z.$$

Thus, $d_{12} = 0$ follows from $Ad_u^*\iota Z = \iota Z$.

(4)
$$d_{1\star} = -<\iota\delta\phi, \Pi_*^r(u)\Pi^r(\varphi_u^-(hgk^{-1})D\psi>$$
$$- <\iota\delta\phi, Ad^*_{g^{-1}h^{-1}}Ad_{\varphi^+_{hg}(u)^{-1}}Ad_k^* Ad^*_{\varphi_u^-(hgk^{-1})}D\psi> + <\iota\delta\phi, D\psi>.$$

which, upon using $\varphi_u^-(hgk^{-1}) = \varphi_u^-(hg)k^{-1}$ and the multiplicativity

$$\Pi^r(\varphi_u^-(hg)k^{-1}) = \Pi^r(\varphi_u^-(hg)),$$

becomes

$$d_{1\star} = <\iota\delta\phi, \big(-\Pi_*^r(u)\Pi^r(\varphi_u^-(hg)) - Ad^*_{(hg)^{-1}}Ad^*_{\varphi_{hg}^+(u)^{-1}}Ad^*_{\varphi_u^-(hg)} + Id\big)D\psi>.$$

But by taking the derivative at $t=0$ in the identity

$$\varphi_u^-(e^{tX}x) = \varphi_u^-(e^{tX})\varphi_{\varphi_{e^{tX}}^+(u)}^-(x)$$

and dualizing immediately yields

$$Id = Ad_u Ad^*_{x^{-1}} Ad_{\varphi_x^+(u)^{-1}} Ad^*_{\varphi_u^-(x)} + \Pi_*^r(u)\Pi^r(\varphi_u^-(x)). \tag{E.3}$$

Thus $d_{1\star} = 0$ again follows from the triviality of $Ad^*_{G^*}$ on $\mathfrak{h}$.

(5)

$$d_{2\star} = <\iota\delta\phi, \big(-Ad^*_{\varphi_u^-(hgk^{-1})}\Pi_*^r(u)\Pi^r(\varphi_u^-(hgk^{-1}))Ad^*_{\varphi_u^-(hgk^{-1})^{-1}}$$
$$+ Ad^*_{k^{-1}}\Pi_*^l(\varphi_{hg}^+(u))\Pi^l(g)Ad_{\varphi_{hg}^+(u)^{-1}}Ad^*_k\big)D'\psi>.$$

Now, by taking the derivative at $t=0$ in the identity

$$k\varphi_v^-(e^{tX})k^{-1} = \varphi_{\varphi_{k^{-1}}^+(v)}^-(ke^{tX}k^{-1})$$

and dualizing, we find

$$Ad_{v^{-1}}Ad_k^* = Ad_k^* Ad_{(\varphi_{k^{-1}}^+ v)^{-1}}. \tag{E.4}$$

Therefore,

$$d_{2\star} = <\iota\delta\phi, \big(-Ad^*_{\varphi_u^-(hgk^{-1})}\Pi_*^r(u)\Pi^r(\varphi_u^-(hgk^{-1}))Ad^*_{\varphi_u^-(hgk^{-1})^{-1}}$$
$$+ \Pi_*^l(\varphi_{hgk^{-1}}^+(u))\Pi^l(hgk^{-1})Ad_{\varphi_{hgk^{-1}}^+(u)^{-1}}\big)D'\psi>,$$

where we have also used the identities (E.1) and (E.2) above.

Next, by taking the derivative at $t=0$ of the expression

$$\varphi_x^+(ue^{t\Lambda}) = \varphi_{\varphi_{e^{t\Lambda}}^-(x)}^+(u)\varphi_x^+(e^{t\Lambda}),$$

we obtain

$$Id = Ad_{\varphi_x^+(u)^{-1}}Ad^*_{\varphi_u^-(x)}Ad_u Ad^*_{x^{-1}} + \Pi_*^l(\varphi_x^+(u))\Pi^l(x). \tag{E.5}$$

Combining $(E.3)$ with $(E.5)$ then yields

$$Ad^*_{\varphi_u^-(x)}\Pi_*^r(u)\Pi^r(\varphi_u^-(x))Ad^*_{\varphi_u^-(x)^{-1}} = Ad^*_{\varphi_x^+(u)}\Pi_*^l(\varphi_x^+(u))\Pi^l(x)Ad_{\varphi_x^+(u)^{-1}}.$$

Thus $Ad_u^* Z = Z$ implies $d_{2\star} = 0$.

(6)

$$d_{\star\star} =$$
$$- <D'\phi, Ad_{\varphi_u^-(hgk^{-1})^{-1}}\Pi^r(\varphi_u^-(hgk^{-1}))\Pi_*^r(u)\Pi^r(\varphi_u^-(hgk^{-1}))Ad^*_{\varphi_u^-(hgk^{-1})^{-1}}D'\psi>$$
$$+ <D'\phi, \big(-Ad_k Ad^*_{\varphi_{hg}^+(u)^{-1}}\Pi^l(g)Ad_{\varphi_{hg}^+(u)^{-1}}Ad_k^* + \Pi^l(\varphi_u^-(hgk^{-1}))\big)D'\psi>.$$

Here we use the Poisson property of $\varphi^- : \overline{G^*} \times G \longrightarrow G$ which may be expressed in the form

$$Ad^*_{\varphi^-_u(x)^{-1}} \Pi^r(\varphi^-_u(x)) \Pi^r_*(u) \Pi^r(\varphi^-_u(x)) Ad^*_{\varphi^-_u(x)^{-1}}$$
$$= -Ad^*_{\varphi^+_x(u)^{-1}} \Pi^l(x) Ad_{\varphi^+_x(u)^{-1}} + \Pi^l(\varphi^-_u(x)).$$

Therefore,

$$d_{\star\star} = \; < D'\phi, Ad^*_{\varphi^+_{hgk^{-1}}(u)^{-1}} \Pi^l(hgk^{-1}) Ad_{\varphi^+_{hgk^{-1}}(u)^{-1}} D'\psi >$$
$$- < D'\phi, Ad_k Ad^*_{\varphi^+_{hg}(u)^{-1}} \Pi^l(g) Ad_{\varphi^+_{hg}(u)^{-1}} Ad^*_k D'\psi > .$$

Thus, $d_{\star\star} = 0$ follows from the identities (E.2) and (E.4).

This concludes the verification that $lhs(P) - rhs(P) = 0$.

(b) This follows from a general result of Weinstein according to which the symplectic leaf passing through (p, g, q) of the base X in the full symplectic dual pair

$$\begin{array}{ccc} & \mathcal{S} & \\ \tilde{\beta}_{\mathcal{H}} \swarrow & & \searrow \tilde{\alpha}_{\mathcal{H}} \\ X & & X \end{array}$$

of Theorem 5.2.13 is given by $\tilde{\alpha}_{\mathcal{H}}(\tilde{\beta}_{\mathcal{H}}^{-1}(p, g, q))$. ∎

Remark 5.2.30. It is instructive to compare the infinitesimal generators of the action $\tilde{\Psi}^-$ with Hamiltonian vectorfields on X. Indeed, for $Z_1, Z_2 \in \mathfrak{h}$ and $\gamma \in \mathfrak{g}^*$, an easy calculation shows that

$$\widetilde{(Z_1, Z_2, \gamma)}(p, g, q) = \Pi^\#_X(Z_2, -T^*_g r_{g^{-1}} \gamma, -Z_1)(p, g, q).$$

To conclude the paper we give two corollaries of Theorem 5.2.29 and a remark on a related result in [**HM**].

First, for the special case when $R = 0$, we have $G^* = \mathfrak{g}^*$, $I^* = \iota^* : \mathfrak{g}^* \to \mathfrak{h}^*$, $\varphi^+_g = Ad^*_g$, and $\varphi^-_u = Id$. Therefore, the Poisson Lie group structure on $H \times (H \ltimes \mathfrak{g}^*)$ is given by

$$(k, (h, A)) \cdot (k', (h', A')) = (kk', (hh', A + Ad^*_{h^{-1}} A'))$$
$$\{f, g\}_{H \times (H \ltimes \mathfrak{g}^*)}(k, (h, A)) = < A, [\delta f, \delta g] >,$$

and the group action becomes

$$\tilde{\psi}^-_0 : H \times (H \ltimes \mathfrak{g}^*) \times X \longrightarrow X$$
$$((k, (h, A)), (p, g, q)) \mapsto (Ad^*_{h^{-1}} p + \iota^*(A), hgk^{-1}, Ad^*_{k^{-1}} q + \iota^*(Ad^*_{hgk^{-1}} A)).$$

Corollary 5.2.31. (Symplectic leaves for $R = 0$.)

(a) $\tilde{\psi}^-_0$ is a left Poisson Lie group action.

(b) The symplectic leaf $\mathcal{L}_{(p,g,q)}$ in $(X, \{\,,\,\}_X)$ passing through the point (p, g, q) is the orbit of (p, g, q) under the action $\tilde{\psi}^-_0$. ∎

Next, we consider the symplectic foliation of a Poisson quotient which we now introduce. Recall from Theorem 2.1.4 (b) that X has a pair of Hamiltonian $H-$ actions with α and β as momentum maps respectively. Combining the two actions, we obtain the action

$$h \cdot (p, g, q) = (Ad^*_{h^{-1}} p, hgh^{-1}, Ad^*_{h^{-1}} q) \tag{5.2.15}$$

which is also Hamiltonian and its equivariant momentum map is given by $J = \alpha - \beta$. Now, $0 \in \mathfrak{h}^*$ is clearly a regular value of J and the corresponding isotropy subgroup is H. Hence it follows from Poisson reduction [**MR**] that $J^{-1}(0)/H$ inherits a Poisson structure $\{\,,\,\}_{J^{-1}(0)/H}$ satisfying

$$\{f_1, f_2\}_{J^{-1}(0)/H} \circ \pi = \{\tilde{f}_1, \tilde{f}_2\}_X \circ i. \tag{5.2.16}$$

Here, $i : J^{-1}(0) \to X$ is the inclusion map, $\pi : J^{-1}(0) \to J^{-1}(0)/H$ is the canonical projection, $f_1, f_2 \in C^\infty(J^{-1}(0)/H)$, and $\tilde{f}_1, \tilde{f}_2$ are (locally defined) smooth extensions of $\pi^* f_1, \pi^* f_2$ with differentials vanishing on the tangent spaces of the $H-$ orbits.

Corollary 5.2.32. *The symplectic leaves of $(J^{-1}(0)/H, \{\,,\,\}_{J^{-1}(0)/H})$ are given by the connected components of $\left(\mathcal{L}_{(p,g,q)} \bigcap J^{-1}(0)\right)/H$, $(p, g, q) \in X$.*

Proof. This is a consequence of the theorem and a result in [**MR**], as the triple $(X, J^{-1}(0), E)$ is Poisson reducible, where E is the tangent space to the $H-$ orbits of the action in (5.2.15). ∎

Clearly, the symplectic leaves of the quotient $G/H \times U$ in Proposition 3.2.5 can also be obtained in a similar way.

Remark 5.2.33. We describe here a related result of Hurtubise and Markman in [**HM**]. Let LG be the loop group of a complex reductive Lie group G, and consider the coboundary dynamical Poisson groupoid $U \times LG \times U$ associated with Felder's elliptic dynamical r-matrix. In [**HM**], the authors restricted their attention to the submanifold $[U \times LG \times U](\Sigma)$, consisting of $(p, g, q) \in U \times LG \times U$ where g is meromorphic over the associated elliptic curve Σ. Then as in (5.2.15), the maximal torus T acts on $[U \times LG \times U](\Sigma)$ (the action on the factors U is now trivial). In this case, the reduced structure on $\bigl([U \times LG \times U](\Sigma) \cap J^{-1}(0)\bigr)/T$ is invariant under the action of the affine Weyl group, and therefore descends to the quotient. It turns out that the quotient is partially compactified by the algebro-geometric moduli $M_\Sigma(G, 0)$ and the Poisson structure extends to the whole of $M_\Sigma(G, 0)$. More generally, it is a main result of [**HM**] that the moduli space $M_\Sigma(G, c)$ (of pairs (P, φ), where P is a principal G- bundle over Σ, of topological type c, and φ is a meromorphic section of its adjoint group bundle) of arbitrary topological type c (of which $c = 0$ is a special case) admits an algebraic Poisson structure. The symplectic leaves of $M_\Sigma(G, c)$ were described in [**HM**] using algebro-geometric methods and were shown to carry algebraically completely integrable systems. It would be interesting to understand this result from our point of view.

Appendix

A.1. Proof of Proposition 2.2.3

The most general bivector field on X is of the form

$$\Pi(df, dg)(p, x, q) = K_1(\delta_1 f, \delta_1 g) + K_2(\delta_2 f, \delta_2 g) + R(\delta_1 f, \delta_2 g)$$
$$- R(\delta_1 g, \delta_2 f) + P_1(\delta_1 f, \partial g) - P_1(\delta_1 g, \partial f)$$
$$+ P_2(\delta_2 f, \partial g) - P_2(\delta_2 g, \partial f) + P_G(\partial f, \partial g),$$

where K_i, R, P_i, P_G are evaluated at (p, x, q).

Set

$$\Omega_{(\omega, Z_1, Z_2, Z_3)} = \big((Z_1, \omega, Z_2), (-Z_2, T_y^*(r_{y^{-1}} \circ l_x)\omega, Z_3), (-Z_1, -T_{xy}^* r_{y^{-1}}\omega, -Z_3)\big),$$

and denote $\big(\Pi \oplus \Pi \oplus -\Pi\big)\big((p, x, q), (q, y, r), (p, xy, r)\big)$ by Π_m. Fix a reference point $q_0 \in U$. We have

$$\Pi_m(\Omega_{(0,Z,0,0)}, \Omega_{(0,0,Z',0)}) = 0 \Leftrightarrow R = 0$$
$$\Pi_m(\Omega_{(0,Z,0,0)}, \Omega_{(0,Z',0,0)}) = 0 \Leftrightarrow K_1(p, x, q) = K_1(p, 1, q_0) =: K(p)$$
$$\Pi_m(\Omega_{(0,0,Z,0)}, \Omega_{(0,0,Z',0)}) = 0 \Leftrightarrow K_2(p, x, q) = -K(q).$$

Now,

$$\Pi_m(\Omega_{(\omega,0,0,0)}, \Omega_{(0,Z',0,0)}) = 0 \Leftrightarrow P_1(p, x, q)(Z', \omega) = P_1(p, xy, r)(Z', T_{xy}^* r_{y^{-1}}\omega).$$

Setting successively $y = 1, r = q_0$, and $x = 1, \omega = T_1^* r_y \omega'$ in the latter equality yields

$$P_1(p, y, r)(Z', \omega') = P_1(p, 1, q_0)(Z', T_1^* r_y \omega') =: <A_1(p)Z', T_1^* r_y \omega'>.$$

Similarly

$$\Pi_m(\Omega_{(\omega,0,0,0)}, \Omega_{(0,0,0,Z')}) = 0 \Leftrightarrow$$
$$P_2(p, x, r)(Z', \omega) = P_2(q_0, 1, r)(Z', T_1^* l_x \omega) =: <A_2(r)Z', T_1^* l_x \omega>.$$

Moreover,
$$\Pi_m(\Omega_{(\omega,0,0,0)}, \Omega_{(0,0,Z',0)}) = 0 \Leftrightarrow A_1(p) = A_2(p).$$

It only remains to demand that $\Pi_m(\Omega_{(\omega,0,0,0)}, \Omega_{(\omega',0,0,0)}) = 0$. But working in the right invariant frame $P_G(\omega, \omega') = <T_1^* r_x \omega, P(T_1^* r_x \omega')>$, the latter condition is equivalent to the cocycle property

$$P(p, xy, r) = P(p, x, q) + Ad_x P(q, y, r) Ad_x^*.$$

Hence the assertion. ∎

A.2. Proof of Theorem 2.2.5 (b)

We have to check the Jacobi identity for the bracket

$$\{f,g\}_X(p,x,q) = <p,[\delta_1 f,\delta_1 g]> - <q,[\delta_2 f,\delta_2 g]>$$
$$- <A_\chi(p)\delta_1 f, Dg> - <A_\chi(q)\delta_2 f, D'g>$$
$$+ <A_\chi(p)\delta_1 g, Df> + <A_\chi(q)\delta_2 g, D'f>$$
$$+ <Df, P(p,x,q)Dg>.$$

We will use (up to sign) the same notation J_{ijk} as in the text. If $a \in \mathfrak{h}$, we define (as in [EV]) the functions $a_1, a_2 \in C^\infty(U \times G \times U)$ by $a_1(p,h,q) = <p,a>$ and $a_2(p,h,q) = <q,a>$. Finally, for $Y \in \mathfrak{g}$, the left (resp. right) invariant vector field on G whose value at 1 is Y will be denoted by Y^l (resp. Y^r).

We now compute J_{ijk} for the various cases.

First of all, it is clear that $J_{ijk} = 0$, $i,j,k \in \{1,2\}$.

On the other hand, we have

$$J_{*12} = \{\{p_G^* f, a_1\}, b_2\} + \{\{a_1, b_2\}, p_G^* f\} + \{\{b_2, p_G^* f\}, a_1\}$$
$$= (A_\chi(q)(b))^l (A_\chi(p)(a))^r(f)(x) - (A_\chi(p)(a))^r(A_\chi(q)(b))^l(f)(x) = 0.$$

while

$$J_{*11} = \{\{p_G^* f, a_1\}, b_1\} + \{\{a_1, b_1\}, p_G^* f\} + \{\{b_1, p_G^* f\}, a_1\}$$
$$= - <dA_\chi(p) \cdot ad_b^* p \cdot a, Df(x)> + (A_\chi(p)b)^l(A_\chi(p)a)^l(f)(x)$$
$$- <A_\chi(p)[a,b], Df(x)> + <dA_\chi(p) \cdot ad_a^* p \cdot b, Df(x)>$$
$$- (A_\chi(p)(a))^l(A_\chi(p)b)^l(f)(x)$$
$$= <dA_\chi(p) \cdot ad_a^* p \cdot b - dA_\chi(p) \cdot ad_b^* p \cdot a$$
$$+ [A_\chi(p)a, A_\chi(p)b] - A_\chi(p)([a,b]), Df(x)>.$$

Similarly,

$$J_{*22} = <-dA_\chi(p) \cdot ad_a^* p \cdot b + dA_\chi(p) \cdot ad_b^* p \cdot a$$
$$- [A_\chi(p)a, A_\chi(p)b] + A_\chi(p)([a,b]), D'f(x)>.$$

So $J_{*ij} = 0$, $i,j \in \{1,2\} \Leftrightarrow A_\chi : \mathfrak{h}^* \times \mathfrak{h} \longrightarrow \mathfrak{g}$ is a morphism of Lie algebroids.

Now,

$$J_{1**} = \{a_1, \{p_G^* f, p_G^* g\}\} + \{p_G^* f, \{p_G^* g, a_1\}\} + \{p_G^* g, \{a_1, p_G^* f\}\}$$
$$= <Df, \delta_1 P \cdot ad_a^* p \cdot Dg> - (A_\chi(p)a)^r(PDg)^r(f)$$
$$- <Df, DP \cdot A_\chi(p)(a) \cdot Dg> + (A_\chi(p)a)^r(PDf)^r(g)$$
$$+ <dA_\chi(p) \cdot (A_\chi(p)^* Df) \cdot a, Dg> - (PDf)^r(A_\chi(p)a)^r(g)$$
$$- <dA_\chi(p) \cdot (A_\chi(p)^* Dg) \cdot a, Df> + (PDg)^r(A_\chi(p)a)^r(f)$$
$$= <Df, \delta_1 P \cdot ad_a^* p \cdot Dg + ad_{A_\chi(p)a}^*(P(Dg))>$$
$$+ <Df, P(ad_{A_\chi(p)a}^* Dg) - DP \cdot (A_\chi(p)a) \cdot Dg>$$
$$+ <dA_\chi(p) \cdot (A_\chi(p)^* Df) \cdot a, Dg> - <dA_\chi(p) \cdot (A_\chi(p)^* Dg) \cdot a, Df>.$$

Similarly,

$$J_{2**} = < Df, -\delta_2 P \cdot ad_a^* q \cdot Dg - D'P \cdot (A_\chi(q)(a)) \cdot Dg >$$
$$+ < dA_\chi(q) \cdot (A_\chi(q)^* D'f) \cdot a, D'g > - < dA_\chi(q) \cdot (A_\chi(q)^* D'g) \cdot a, D'f > .$$

Writing the groupoid 1- cocycle as

$$P(p, x, q) = -l(p) + \pi(x) + Ad_x l(q) Ad_x^*,$$

we have

$$\delta_1 P \cdot \Lambda = -dl(p) \cdot \Lambda$$
$$DP \cdot X = d\pi(1)(X) + ad_X \pi(x) + \pi(x) ad_X^*$$
$$+ ad_X Ad_x l(q) Ad_x^* + Ad_x l(q) Ad_x^* ad_X^*$$
$$\delta_2 P \cdot \Lambda = Ad_x dl(q) \cdot \Lambda Ad_x^*$$
$$D'P \cdot X = Ad_x d\pi(1) \cdot X Ad_x^* + Ad_x ad_X l(q) Ad_x^*$$
$$+ Ad_x l(q) ad_X^* Ad_x^*$$

Inserting the latter into J_{1**} and J_{2**} yields

$$J_{**1} = 0 \Leftrightarrow J_{**2} = 0 \Leftrightarrow$$
$$< \alpha, (dl(p) ad_a^* p + ad_{A_\chi(p)a} l(p) + l(p) ad_{A_\chi(p)a}^* + d\pi(1) A_\chi(p) a) \beta >$$
$$= + < dA_\chi(p) \cdot (A_\chi^*(p) \alpha) \cdot a, \beta > - < dA_\chi(p) \cdot (A_\chi(p)^* \beta) \cdot a, \alpha >,$$
$$\forall \alpha, \beta \in \mathfrak{g}^*, a \in \mathfrak{h}.$$

Finally,

$$J_{***} = \{p_G^* f, \{p_G^* g, p_G^* h\}\} + c.p.(f, g, h)$$
$$= < Dg, \delta_1 P \cdot (A_\chi(p)^* Df) \cdot Dh + \delta_2 P \cdot (A_\chi(p)^* D'f) \cdot Dh >$$
$$- (PDf)^r (PDh)^r (g) + (PDf)^r (PDg)^r (h)$$
$$- < Dg, DP \cdot (PDf) \cdot Dh > + c.p.(f, g, h),$$

which may easily be brought to the form stated in Theorem 2.2.5 (b). ∎

A.3. Proof of Proposition 3.2.1

We use the same notations as in the text. We briefly indicate the main step of the calculation of the Lie bracket of 1− forms

$$[\omega, \omega'](q, 1, q) := -L_{\Pi^\# \omega} \omega' + L_{\Pi^\# \omega'} \omega - d < \omega, \Pi^\# \omega' > .$$

Writing

$$< L_{\Pi^\# \omega} \omega', (\Lambda, X^l, \Lambda') > = d < \omega', (\Lambda, X^l, \Lambda' > \cdot \Pi^\# \omega$$
$$+ < \omega', [(\Lambda, X^l, \Lambda'), \Pi^\# \omega] >$$
$$= I + II$$

we have

$$I = d<\omega', (\Lambda, X^l, \Lambda')> \cdot \Pi^\#(q,1,q)\omega$$
$$= d<(-Z', B'^l, Z'), (\Lambda, X^l, \Lambda')> \cdot \Pi^\#(q,1,q)(-Z(q), B(q), Z(q))$$
$$= d<(-Z', B', Z'), (\Lambda, X, \Lambda')> \cdot$$
$$(K(q)Z(q) - A^*(q)B(q), 0, K(q)Z(q) - A^*(q)B(q))$$
$$= -<dZ'(q)(K(q)Z(q) - A^*(q)B(q)), \Lambda>$$
$$+ <dB'(K(q)Z(q) - A^*(q)B(q)), X> + <dZ'(q)(K(q)Z(q) - A^*(q)B(q)), \Lambda'>.$$

Consider the flow $\phi_t(p, x, q) = (p + t\Lambda, xe^{tX}, q + t\Lambda')$ of the vector field (Λ, X^l, Λ') and set $\Pi^\#\omega(p, x, q) = (\pi_1, \pi_\star, \pi_2) \in h^* \times T_xG \times h^*$. We have

$$II = <\omega', [(\Lambda, X^l, \Lambda'), \Pi^\#\omega]>(q,1,q)$$
$$= <\omega'(q,1,q), \frac{d}{dt}\bigg|_0 \left(T_{\phi_t(q,1,q)}\phi_{-t}\Pi^\#\omega(\phi_t(q,1,q))\right)>$$
$$= <\omega'(q,1,q), \frac{d}{dt}\bigg|_0 \left(\pi_1(\phi_t(q,1,q)), T_{e^{tx}}r_{e^{-tx}}\pi_\star(\phi_t(q,1,q)), \pi_2(\phi_t(q,1,q)))\right)>$$
$$= <(-Z'(q), B'(q), Z'(q)),$$
$$\frac{d}{dt}\bigg|_0 \big(K(q+t\Lambda)Z(q+t\Lambda') - A^*(q+t\Lambda)Ad^*_{e^{-tx}}B(q+t\Lambda'),$$
$$- A(q+t\Lambda)Z(q+t\Lambda') + Ad_{e^{tx}}A(q+t\Lambda')Z(q+t\Lambda')$$
$$+ P(q+t\Lambda, e^{tX}, q+t\Lambda')Ad^*_{e^{-tx}}B(q+t\Lambda'),$$
$$K(q+t\Lambda')Z(q+t\Lambda') - A^*(q+t\Lambda')B(q+t\Lambda'))>.$$

On the other hand,

$$<d<\omega, \Pi^\#\omega'>, (\Lambda, X^l, \Lambda')>(q,1,q)$$
$$= \frac{d}{dt}\bigg|_0 <\omega, \Pi^\#\omega'>(q+t\Lambda, e^{tX}, q+t\Lambda')$$
$$= \frac{d}{dt}\bigg|_0 <(-Z(q+t\Lambda'), T^*_{e^{tx}}l_{e^{-tx}}B(q+t\Lambda'), Z(q+t\Lambda')),$$
$$\Pi^\#(q+t\Lambda, e^{tX}, q+t\Lambda')(-Z'(q+t\Lambda'), T^*_{e^{tx}}l_{e^{-tx}}B'(q+t\Lambda'), Z'(q+t\Lambda'))>$$
$$= \frac{d}{dt}\bigg|_0 \big(<-Z(q+t\Lambda'), K(q+t\Lambda)Z'(q+t\Lambda') - A^*(q+t\Lambda)Ad^*_{e^{-tx}}B'(q+t\Lambda')>$$
$$+ <B(q+t\Lambda'), -Ad_{e^{-tx}}A(q+t\Lambda)Z'(q+t\Lambda') + A(q+t\Lambda')Z'(q+t\Lambda')$$
$$+ Ad_{e^{-tx}}P(q+t\Lambda, e^{tX}, q+t\Lambda')Ad^*_{e^{-tx}}B'(q+t\Lambda')>$$
$$+ <Z(q+t\Lambda'), K(q+t\Lambda')Z'(q+t\Lambda') - A^*(q+t\Lambda')B'(q+t\Lambda')> \big).$$

Collecting terms and expanding out the derivatives, one easily obtains, after the identification ι_-, the asserted Lie bracket.

A.4. Proof of the coisotropy in Theorem 5.1.4

We have to show that the graph of the multiplication
$$Gr(m) \subset \Gamma \times \Gamma \times \overline{\Gamma}$$
is a coisotropic submanifold. We use (as in (5.1.6)) the notation $\phi_h^l(v) = \phi_v^l(h) = \varphi_{h^{-1}}^+(v), h \in H, v \in G^*$. We have

$Gr(m) =$
$$\{((h, Ad^*_{k^{-1}}q + I^*(v), u), (k, q, v), (hk, q, u\phi_h^l(v))) \mid h, k \in H, u, v \in G^*, q \in \mathfrak{h}^*\},$$
therefore
$$T_*(Gr(m)) = \{((Z_1, -Ad^*_{k^{-1}} ad^*_{(T_k l_{k^{-1}} Z_2)}q + Ad^*_{k^{-1}}\lambda + T_v(I^*)V, U),$$
$$(Z_2, \lambda, V), (T_k l_h Z_2 + T_h r_k Z_1, \lambda, T_u r_{\phi_h^l(v)} U + T_{\phi_h^l(v)} l_u (T_h \phi_v^l Z_1 + T_v \phi_h^l V))$$
$$\mid Z_1 \in T_h H, Z_2 \in T_k H, \lambda \in \mathfrak{h}^*, U \in T_u G^*, V \in T_v G^*\}.$$

Hence $\Omega \in (T_* Gr(m))^\perp$ if and only if
$$\Omega = ((-T_h^* r_k \mu - T_h^*(l_u \circ \phi_v^l)A, z_1, -T_u^* r_{\phi_h^l(v)} A),$$
$$(-T_k^* l_h \mu - T_k^* l_{k^{-1}} ad^*_{(Ad_{k^{-1}} z_1)}q, z_2, -T_v^* I^* z_1 - T_v^*(l_u \circ \phi_h^l)A),$$
$$(\mu, -Ad_{k^{-1}} z_1 - z_2, A)),$$
for some $\mu \in T^*_{hk} H, z_1, z_2 \in \mathfrak{h}, A \in T^*_{u\phi_h^l(v)} G^*$. For $\Omega, \Omega' \in (T_* Gr(m))^\perp$, we have

$(\Pi \oplus \Pi \oplus -\Pi)(\Omega, \Omega') = <T_h^* r_k \mu' + T_h^*(l_u \circ \phi_v^l)A', T_1 l_h z_1>$
$- <T_h^* r_k \mu + T_h^*(l_u \circ \phi_v^l)A, T_1 l_h z_1'> - <Ad^*_{k^{-1}}q + I^*(v), [z_1, z_1']>$
$- <T_u^* r_{\phi_h^l(v)} A, \lambda^+(T_1^* l_u (T_u^* r_{\phi_h^l(v)} A'))(u)>$
$+ <T_k^* l_h \mu' + T_k^* l_{k^{-1}} ad^*_{(Ad_{k^{-1}} z_1')}q, T_1 l_k z_2>$
$- <T_k^* l_h \mu + T_k^* l_{k^{-1}} ad^*_{(Ad_{k^{-1}} z_1)}q, T_1 l_k z_2'> - <q, [z_1, z_1']>$
$- <T_v^* I^* z_1 + T_v^*(l_u \circ \phi_h^l)A, \lambda^+(T_1^* l_v (T_v^* I^* z_1' + T_v^*(l_u \circ \phi_h^l)A'))(v)>$
$- <\mu', T_1 l_{hk}(Ad_{k^{-1}} z_1 + z_2)> + <\mu, T_1 l_{hk}(Ad_{k^{-1}} z_1' + z_2')>$
$+ <q, [Ad_{k^{-1}} z_1 + z_2, Ad_{k^{-1}} z_1' + z_2']> + <A, \lambda^+(T_1^* l_{u\phi_h^l(v)} A')(u\phi_h^l(v))>.$

We now treat separately the three types of terms which do not obviously cancel out:

$(1) := - <I^*(v), [z_1, z_1']> - <T_v^* I^* z_1, \lambda^+(T_1^* l_v T_v^* I^* z_1')(v)>.$

Since I^* is a morphism of groups, $I^* \circ l_v(w) = I^*(u) + I^*(w)$ therefore $T_1(I^* \circ l_v) = \iota^*$, and hence

$$T_v I^* \lambda^+(T_1^*(I^* \circ l_v)z_1')(v) = T_v I^* \lambda^+(\iota(z_1'))(v)$$
$$= \frac{d}{dt}\bigg|_0 I^*(\phi^l_{e^{t\iota(z_1')}}(v))$$
$$= \frac{d}{dt}\bigg|_0 Ad^*_{e^{t\iota(z_1')}}(I^*(v)) = ad^*_{z_1'} I^*(v).$$

Thus $(1) = 0$.

$$(2) := - <T_u^* r_{\phi_h^l(v)} A, \lambda^+(T_1^* l_u(T_u^* r_{\phi_h^l(v)} A'))(u)>$$
$$- <T_v^*(l_u \circ \phi_h^l) A, \lambda^+(T_1^* l_v(T_v^*(l_u \circ \phi_h^l) A'))(v)>$$
$$+ <A, \lambda^+(T_1^* l_{u\phi_h^l(v)} A')(u\phi_h^l(v))>.$$

Let Π_* be the Poisson tensor of G^*. We have

$$(2) = \Pi_*(u)(T_u^* r_{\phi_h^l(v)} A, T_u^* r_{\phi_h^l(v)} A')$$
$$+ \Pi_*(v)(T_v^*(l_u \circ \phi_h^l) A, T_v^*(l_u \circ \phi_h^l) A') - \Pi_*(u\phi_h^l(v))(A, A')$$
$$= +\Pi_*(v)(T_v^*(l_u \circ \phi_h^l) A, T_v^*(l_u \circ \phi_h^l) A') - \Pi_*(\phi_h^l(v))(T_{\phi_h^l(v)}^* l_u A, T_{\phi_h^l(v)}^* l_u A'),$$

where in the last equality we have used the multiplicativity of Π_*. Thus $(2) = 0$ follows from the Hamiltonian property of $\phi_h^l : G^* \longrightarrow G^*$.

$$(3) := <T_h^*(l_u \circ \phi_v^l) A', T_1 l_h z_1> - <T_v^* I^* z_1, \lambda^+(T_1^* l_v(T_v^*(l_u \circ \phi_h^l) A'))(v)>.$$

We have

$$<T_h^*(l_u \circ \phi_v^l) A', T_1 l_h z_1> = <A', T_1(l_u \circ \phi_v^l \circ l_h) z_1>$$
$$= <A', T_1(l_u \circ \phi_h^l \circ \phi_v^l) z_1>$$
$$= <T_v^*(l_u \circ \phi_h^l) A', T_1 \phi_v^l z_1>,$$

and $T_v^* I^* = T_v^* l_{v^{-1}} \circ \iota$. Therefore

$$(3) = <T_v^*(l_u \circ \phi_h^l) A', T_1 \phi_v^l(z_1)> + <T_v^* l_{v^{-1}} \iota(z_1), \Pi_*(v)(T_v^*(l_u \circ \phi_h^l) A')>$$
$$= <T_v^*(l_u \circ \phi_h^l) A', T_1 \phi_v^l(z_1) - \Pi_*(v)(T_v^* l_{v^{-1}} \iota(z_1))> = 0.$$

Hence the proof. ∎

Bibliography

[AM] Almeida, R. and Molino, P., *Suites d'Atiyah et feuilletages transversalement complets*, C. R. Acad. Sci. Paris, Serie I, t. 300 (1985), 13-15.

[BD] Belavin, A.A. and Drinfel'd, V., *Triangle equations for simple Lie algebras*, Mathematical Physics Reviews (Ed. Novikov et al.) Harwood, New York (1984), 93-165.

[BDF] Balog, J., Dąbrowski, L., Fehér, L., *Classical r-matrix and exchange algebra in WZNW and Toda theories*, Phys. Lett. B **244** (1990), 227-234.

[BH] Brown, R. and Higgins, P. J., *On the connection between the second relative homotopy groups of some related spaces*, Proc. London Math. Soc. **36** (1978), 193-212.

[BKS] Bangoura, M. and Kosmann-Schwarzbach, Y., *Équations de Yang-Baxter dynamique classique et algébroïdes de Lie*, C. R. Acad. Sc. Paris, Serie I **327** (1998), 541-546.

[CdSW] Cannas da Silva, A. and Weinstein, A., *Geometric models for noncommutative algebras*, Berkeley Mathematics Lecture Notes 10. Amer. Math. Soc., Providence, RI (1999).

[CDW] Coste, A., Dazord, P. and Weinstein, A., *Groupoides symplectiques*, Publications du Départment de Mathématiques de l'Université de Lyon **2/A** (1987), 1-65.

[CF] Crainic, M. and Fernandes, R., *Integrability of Lie brackets*, Ann. of Math. **157** (2003), 575-620.

[D] Drinfel'd, V.G., *Hamiltonian structures on Lie groups, Lie bialgebras, and the geometric meaning of the classical Yang-Baxter equations*, Soviet Math. Dokl. **27** (1983), 68-71.

[E] Ehresmann, C., *Catégories structurées*, Ann. Sci. École Norm. Sup. **80** (1963), 349-426.

[ES] Etingof, P. and Schiffmann, O., *On the moduli space of classical dynamical r-matrices*, Math. Res. Lett. **8 no 1-2** (2001), 157-170.

[EV] Etingof, P. and Varchenko, A., *Geometry and classification of solutions of the classical dynamical Yang-Baxter equation*, Commun. Math. Phys. **192** (1998), 77-120.

[F] Felder, G., *Conformal field theory and integrable systems associated to elliptic curves*, Proc. ICM Zurich, Birkhäuser, Basel (1994), 1247–1255.

[HM] Hurtubise, J., Markman, E., *Elliptic Sklyanin integrable systems for arbitrary reductive groups*, Adv. Theor. Math. Phys. **6** (2002), 873-978.

[K] Karasev, M., *Analogues of objects of the theory of Lie groups for nonlinear Poisson brackets*, Math. USSR Izvestiya **28** (1987), 497-527.

[Ko] Koszul, J.L., *Crochet de Schouten-Nijenhuis et cohomologie*, Astérisque, hors série, Soc. Math. France, Paris (1985), 257-271.

[L] Libermann, P., *On symplectic and contact groupoids*, Differential geometry and its applications (Opava 1992), Math. Publ.1, Silesian Univ. Opava, Opava (1993), 29-45.

[L1] Li, L.-C., *Coboundary dynamical Poisson groupoids and integrable systems*, Int. Math. Res. Not. 2003 **51** (2003), 2725-2746.

[L2] ———, *A family of hyperbolic spin Calogero-Moser systems and the spin Toda lattices*, Comm. Pure Appl.Math. **57** (2004), 791-832.

[Lu] Lu, J.-H., *Ph.D. Thesis, Berkeley*, 1990.

[LW1] Lu, J.-H., Weinstein, A., *Poisson Lie groups, dressing transformations, and Bruhat decompositions.*, J. Diff. Geom. 31 (1990), 501–526.

[LW2] Lu, J.-H., Weinstein, A., *Groupoides symplectiques doubles des groupes de Lie-Poisson*, C. R. Acad. Sci. Paris, Ser. I, t. 309 (1989), 951-954.

[LWX] Liu, Z.-J., Weinstein, A. , and Xu,P., *Manin triples for Lie bialgebroids*, J. Diff. Geom. **45** (1997), 547-574.

[LX1] Li, L.-C. and Xu, P., *Spin Calogero-Moser systems associated with simple Lie algebras*, C. R. Acad. Sci. Paris, t.331, Série I (2000), 55–60.

BIBLIOGRAPHY

[LX2] _____, *A class of integrable spin Calogero-Moser systems*, Commun. Math. Phys. **231** (2002), 257-286.

[LiX] Liu, Z.-J. and Xu, P., *The local structure of Lie bialgebroids*, Lett. Math. Phys. **61** (2002), 15-28.

[M1] Mackenzie, K., *Lie groupoids and Lie algebroids in differential geometry*. LMS Lecture Notes Series *124*, Cambridge University Press, 1987.

[M2] _____, *On symplectic double groupoids and the duality of Poisson groupoids*, Internat. J. Math. **10** (1999), 435-456.

[M3] _____, *Double Lie algebroids and second-order geometry I.*, Adv. Math. **94** (1992), 180-239.

[MM] Magri, F., Morosi, C., *A geometrical characterization of integrable Hamiltonian systems through the theory of Poisson-Nijenhuis manifolds*, Quaderne 519. Dipart. di Mate. Univ. de Milano (1984).

[MR] Marsden J.E. and Ratiu T., *Reduction of Poisson manifolds*, Letters in Math. Phys. **11** (1986), 161-169.

[MW] Mikami K. and Weinstein A., *Moments and reduction for symplectic groupoids*, Publ. RIMS, Kyoto University **24** (1988), 121-140.

[MX1] Mackenzie, K. and Xu, P., *Lie bialgebroids and Poisson groupoids*, Duke Math. J. **73** (1994), 415–452.

[MX2] _____, *Integration of Lie bialgebroids*, Topology **39** (2000), 445-467.

[P] Pradines, J., *Géometrie differentielle au-dessus d'un groupoide*, C. R. Acad. Sc. Paris, Ser. I, t. 266 (1968), 1194-1196.

[STS] Semenov-Tian-Shansky, M., *Dressing transformations and Poisson Lie group actions*, Publ. RIMS, Kyoto University **21** (1985), 1237-1260.

[W1] Weinstein, A., *Coisotropic calculus and Poisson groupoids*, J. Math. Soc. Japan 4 no. 40 (1988), 705–727.

[W2] Weinstein, A., *Symplectic groupoids and Poisson manifolds*, Bull. Amer. Math. Soc. **16** (1987), 101-104.

[X] Xu, P., *On Poisson groupoids*, Internat. J. Math. **6** (1995), 101-124.

Editorial Information

To be published in the *Memoirs*, a paper must be correct, new, nontrivial, and significant. Further, it must be well written and of interest to a substantial number of mathematicians. Piecemeal results, such as an inconclusive step toward an unproved major theorem or a minor variation on a known result, are in general not acceptable for publication. Papers appearing in *Memoirs* are generally longer than those appearing in *Transactions*, which shares the same editorial committee.

As of November 30, 2004, the backlog for this journal was approximately 5 volumes. This estimate is the result of dividing the number of manuscripts for this journal in the Providence office that have not yet gone to the printer on the above date by the average number of monographs per volume over the previous twelve months, reduced by the number of volumes published in four months (the time necessary for preparing a volume for the printer). (There are 6 volumes per year, each containing at least 4 numbers.)

A Consent to Publish and Copyright Agreement is required before a paper will be published in the *Memoirs*. After a paper is accepted for publication, the Providence office will send a Consent to Publish and Copyright Agreement to all authors of the paper. By submitting a paper to the *Memoirs*, authors certify that the results have not been submitted to nor are they under consideration for publication by another journal, conference proceedings, or similar publication.

Information for Authors

Memoirs are printed from camera copy fully prepared by the author. This means that the finished book will look exactly like the copy submitted.

The paper must contain a *descriptive title* and an *abstract* that summarizes the article in language suitable for workers in the general field (algebra, analysis, etc.). The *descriptive title* should be short, but informative; useless or vague phrases such as "some remarks about" or "concerning" should be avoided. The *abstract* should be at least one complete sentence, and at most 300 words. Included with the footnotes to the paper should be the 2000 *Mathematics Subject Classification* representing the primary and secondary subjects of the article. The classifications are accessible from www.ams.org/msc/. The list of classifications is also available in print starting with the 1999 annual index of *Mathematical Reviews*. The Mathematics Subject Classification footnote may be followed by a list of *key words and phrases* describing the subject matter of the article and taken from it. Journal abbreviations used in bibliographies are listed in the latest *Mathematical Reviews* annual index. The series abbreviations are also accessible from www.ams.org/publications/. To help in preparing and verifying references, the AMS offers MR Lookup, a Reference Tool for Linking, at www.ams.org/mrlookup/. When the manuscript is submitted, authors should supply the editor with electronic addresses if available. These will be printed after the postal address at the end of the article.

Electronically prepared manuscripts. The AMS encourages electronically prepared manuscripts, with a strong preference for $\mathcal{AMS}$-LaTeX. To this end, the Society has prepared $\mathcal{AMS}$-LaTeX author packages for each AMS publication. Author packages include instructions for preparing electronic manuscripts, the *AMS Author Handbook*, samples, and a style file that generates the particular design specifications of that publication series. Though $\mathcal{AMS}$-LaTeX is the highly preferred format of TeX, author packages are also available in $\mathcal{AMS}$-TeX.

Authors may retrieve an author package from e-MATH starting from www.ams.org/tex/ or via FTP to ftp.ams.org (login as anonymous, enter username as password, and type cd pub/author-info). The *AMS Author Handbook* and the *Instruction Manual* are available in PDF format following the author packages link from www.ams.org/tex/. The author package can be obtained free of charge by sending email to pub@ams.org (Internet) or from the Publication Division, American Mathematical Society, 201 Charles St., Providence, RI 02904, USA. When requesting an author package, please specify $\mathcal{AMS}$-LaTeX or $\mathcal{AMS}$-TeX, Macintosh or IBM (3.5) format, and the publication in which your paper will appear. Please be sure to include your complete mailing address.

Sending electronic files. After acceptance, the source file(s) should be sent to the Providence office (this includes any TeX source file, any graphics files, and the DVI or PostScript file).

Before sending the source file, be sure you have proofread your paper carefully. The files you send must be the EXACT files used to generate the proof copy that was accepted for publication. For all publications, authors are required to send a printed copy of their paper, which exactly matches the copy approved for publication, along with any graphics that will appear in the paper.

TeX files may be submitted by email, FTP, or on diskette. The DVI file(s) and PostScript files should be submitted only by FTP or on diskette unless they are encoded properly to submit through email. (DVI files are binary and PostScript files tend to be very large.)

Electronically prepared manuscripts can be sent via email to pub-submit@ams.org (Internet). The subject line of the message should include the publication code to identify it as a Memoir. TeX source files, DVI files, and PostScript files can be transferred over the Internet by FTP to the Internet node e-math.ams.org (130.44.1.100).

Electronic graphics. Comprehensive instructions on preparing graphics are available at www.ams.org/jourhtml/graphics.html. A few of the major requirements are given here.

Submit files for graphics as EPS (Encapsulated PostScript) files. This includes graphics originated via a graphics application as well as scanned photographs or other computer-generated images. If this is not possible, TIFF files are acceptable as long as they can be opened in Adobe Photoshop or Illustrator. No matter what method was used to produce the graphic, it is necessary to provide a paper copy to the AMS.

Authors using graphics packages for the creation of electronic art should also avoid the use of any lines thinner than 0.5 points in width. Many graphics packages allow the user to specify a "hairline" for a very thin line. Hairlines often look acceptable when proofed on a typical laser printer. However, when produced on a high-resolution laser imagesetter, hairlines become nearly invisible and will be lost entirely in the final printing process.

Screens should be set to values between 15% and 85%. Screens which fall outside of this range are too light or too dark to print correctly. Variations of screens within a graphic should be no less than 10%.

Inquiries. Any inquiries concerning a paper that has been accepted for publication should be sent directly to the Electronic Prepress Department, American Mathematical Society, 201 Charles St., Providence, RI 02904, USA.

Editors

This journal is designed particularly for long research papers, normally at least 80 pages in length, and groups of cognate papers in pure and applied mathematics. Papers intended for publication in the *Memoirs* should be addressed to one of the following editors. In principle the Memoirs welcomes electronic submissions, and some of the editors, those whose names appear below with an asterisk (*), have indicated that they prefer them. However, editors reserve the right to request hard copies after papers have been submitted electronically. Authors are advised to make preliminary email inquiries to editors about whether they are likely to be able to handle submissions in a particular electronic form.

*Algebra to ROBERT GURALNICK, Department of Mathematics, University of Southern California, Los Angeles, CA 90089-1113; email: guralnic@math.usc.edu

Algebraic geometry to DAN ABRAMOVICH, Department of Mathematics, Boston University, 111 Cummington St., Boston, MA 02215; email: abramovic@bu.edu

*Algebraic number theory to V. KUMAR MURTY, Department of Mathematics, University of Toronto, 100 St. George Street, Toronto, ON M5S 1A1, Canada; email: murty@math.toronto.edu

*Algebraic topology to ALEJANDRO ADEM, Department of Mathematics, University of Wisconsin, 480 Lincoln Drive, Madison, WI 53706-1388; email: adem@math.wisc.edu

Combinatorics and Lie theory to SERGEY FOMIN, Department of Mathematics, University of Michigan, Ann Arbor, Michigan 48109-1109; email: fomin@umich.edu

Complex analysis and complex geometry to DUONG H. PHONG, Department of Mathematics, Columbia University, 2990 Broadway, New York, NY 10027-0029; email: phong@math.columbia.edu

*Differential geometry and global analysis to LISA C. JEFFREY, Department of Mathematics, University of Toronto, 100 St. George St., Toronto, ON Canada M5S 3G3; email: jeffrey@math.toronto.edu

Dynamical systems and ergodic theory to ROBERT F. WILLIAMS, Department of Mathematics, University of Texas, Austin, Texas 78712-1082; email: bob@math.utexas.edu

*Functional analysis and operator algebras to MARIUS DADARLAT, Department of Mathematics, Purdue University, 150 N. University St., West Lafayette, IN 47907-2067; email: mdd@math.purdue.edu

*Geometric analysis to TOBIAS COLDING, Courant Institute, New York University, 251 Mercer St., New York, NY 10012; email: colding@cims.nyu.edu

*Geometric analysis to MLADEN BESTVINA, Department of Mathematics, University of Utah, 155 South 1400 East, JWB 233, Salt Lake City, Utah 84112-0090; email: bestvina@math.utah.edu

Harmonic analysis to ALEXANDER NAGEL, Department of Mathematics, University of Wisconsin, 480 Lincoln Drive, Madison, WI 53706-1313; email: nagel@math.wisc.edu

Harmonic analysis, representation theory, and Lie theory to ROBERT J. STANTON, Department of Mathematics, The Ohio State University, 231 West 18th Avenue, Columbus, OH 43210-1174; email: stanton@math.ohio-state.edu

*Logic to STEFFEN LEMPP, Department of Mathematics, University of Wisconsin, 480 Lincoln Drive, Madison, Wisconsin 53706-1388; email: lempp@math.wisc.edu

Number theory to HAROLD G. DIAMOND, Department of Mathematics, University of Illinois, 1409 W. Green St., Urbana, IL 61801-2917; email: diamond@math.uiuc.edu

*Ordinary differential equations, and applied mathematics to PETER W. BATES, Department of Mathematics, Michigan State University, East Lansing, MI 48824-1027; email: peter@math.msu.edu

*Partial differential equations to PATRICIA E. BAUMAN, Department of Mathematics, Purdue University, West Lafayette, IN 47907-1395; email: bauman@math.purdue.edu

*Probability and statistics to KRZYSZTOF BURDZY, Department of Mathematics, University of Washington, Box 354350, Seattle, Washington 98195-4350; email: burdzy@math.washington.edu

*Real analysis and partial differential equations to DANIEL TATARU, Department of Mathematics, University of California, Berkeley, Berkeley, CA 94720; email: tataru@math.berkeley.edu

All other communications to the editors should be addressed to the Managing Editor, WILLIAM BECKNER, Department of Mathematics, University of Texas, Austin, TX 78712-1082; email: beckner@math.utexas.edu.

Titles in This Series

824 **Luen-Chau Li and Serge Parmentier,** On dynamical Poisson groupoids I, 2005

823 **Claus Mokler,** An analogue of a reductive algebraic monoid whose unit group is a Kac-Moody group, 2005

822 **Stefano Pigola, Marco Rigoli, and Alberto G. Setti,** Maximum principles on Riemannian manifolds and applications, 2005

821 **Nicole Bopp and Hubert Rubenthaler,** Local zeta functions attached to the minimal spherical series for a class of symmetric spaces, 2005

820 **Vadim A. Kaimanovich and Mikhail Lyubich,** Conformal and harmonic measures on laminations associated with rational maps, 2005

819 **F. Andreatta and E. Z. Goren,** Hilbert modular forms: Mod p and p-adic aspects, 2005

818 **Tom De Medts,** An algebraic structure for Moufang quadrangles, 2005

817 **Javier Fernández de Bobadilla,** Moduli spaces of polynomials in two variables, 2005

816 **Francis Clarke,** Necessary conditions in dynamic optimization, 2005

815 **Martin Bendersky and Donald M. Davis,** V_1-periodic homotopy groups of $SO(n)$, 2004

814 **Johannes Huebschmann,** Kähler spaces, nilpotent orbits, and singular reduction, 2004

813 **Jeff Groah and Blake Temple,** Shock-wave solutions of the Einstein equations with perfect fluid sources: Existence and consistency by a locally inertial Glimm scheme, 2004

812 **Richard D. Canary and Darryl McCullough,** Homotopy equivalences of 3-manifolds and deformation theory of Kleinian groups, 2004

811 **Ottmar Loos and Erhard Neher,** Locally finite root systems, 2004

810 **W. N. Everitt and L. Markus,** Infinite dimensional complex symplectic spaces, 2004

809 **J. T. Cox, D. A. Dawson, and A. Greven,** Mutually catalytic super branching random walks: Large finite systems and renormalization analysis, 2004

808 **Hagen Meltzer,** Exceptional vector bundles, tilting sheaves and tilting complexes for weighted projective lines, 2004

807 **Carlos A. Cabrelli, Christopher Heil, and Ursula M. Molter,** Self-similarity and multiwavelets in higher dimensions, 2004

806 **Spiros A. Argyros and Andreas Tolias,** Methods in the theory of hereditarily indecomposable Banach spaces, 2004

805 **Philip L. Bowers and Kenneth Stephenson,** Uniformizing dessins and Belyĭ maps via circle packing, 2004

804 **A. Yu Ol'shanskii and M. V. Sapir,** The conjugacy problem and Higman embeddings, 2004

803 **Michael Field and Matthew Nicol,** Ergodic theory of equivariant diffeomorphisms: Markov partitions and stable ergodicity, 2004

802 **Martin W. Liebeck and Gary M. Seitz,** The maximal subgroups of positive dimension in exceptional algebraic groups, 2004

801 **Fabio Ancona and Andrea Marson,** Well-posedness for general 2×2 systems of conservation law, 2004

800 **V. Poénaru and C. Tanasi,** Equivariant, almost-arborescent representation of open simply-connected 3-manifolds; A finiteness result, 2004

799 **Barry Mazur and Karl Rubin,** Kolyvagin systems, 2004

798 **Benoît Mselati,** Classification and probabilistic representation of the positive solutions of a semilinear elliptic equation, 2004

797 **Ola Bratteli, Palle E. T. Jorgensen, and Vasyl' Ostrovs'kyĭ,** Representation theory and numerical AF-invariants, 2004

796 **Marc A. Rieffel,** Gromov-Hausdorff distance for quantum metric spaces/Matrix algebras converge to the sphere for quantum Gromov-Hausdorff distance, 2004

TITLES IN THIS SERIES

795 **Adam Nyman,** Points on quantum projectivizations, 2004
794 **Kevin K. Ferland and L. Gaunce Lewis, Jr.,** The $RO(G)$-graded equivariant ordinary homology of G-cell complexes with even-dimensional cells for $G = \mathbb{Z}/p$, 2004
793 **Jindřich Zapletal,** Descriptive set theory and definable forcing, 2004
792 **Inmaculada Baldomá and Ernest Fontich,** Exponentially small splitting of invariant manifolds of parabolic points, 2004
791 **Eva A. Gallardo-Gutiérrez and Alfonso Montes-Rodríguez,** The role of the spectrum in the cyclic behavior of composition operators, 2004
790 **Thierry Lévy,** Yang-Mills measure on compact surfaces, 2003
789 **Helge Glöckner,** Positive definite functions on infinite-dimensional convex cones, 2003
788 **Robert Denk, Matthias Hieber, and Jan Prüss,** $\mathcal{R}$-boundedness, Fourier multipliers and problems of elliptic and parabolic type, 2003
787 **Michael Cwikel, Per G. Nilsson, and Gideon Schechtman,** Interpolation of weighted Banach lattices/A characterization of relatively decomposable Banach lattices, 2003
786 **Arnd Scheel,** Radially symmetric patterns of reaction-diffusion systems, 2003
785 **R. R. Bruner and J. P. C. Greenlees,** The connective K-theory of finite groups, 2003
784 **Desmond Sheiham,** Invariants of boundary link cobordism, 2003
783 **Ethan Akin, Mike Hurley, and Judy A. Kennedy,** Dynamics of topologically generic homeomorphisms, 2003
782 **Masaaki Furusawa and Joseph A. Shalika,** On central critical values of the degree four L-functions for GSp(4): The Fundamental Lemma, 2003
781 **Marcin Bownik,** Anisotropic Hardy spaces and wavelets, 2003
780 **S. Marmi and D. Sauzin,** Quasianalytic monogenic solutions of a cohomological equation, 2003
779 **Hansjörg Geiges,** h-principles and flexibility in geometry, 2003
778 **David B. Massey,** Numerical control over complex analytic singularities, 2003
777 **Robert Lauter,** Pseudodifferential analysis on conformally compact spaces, 2003
776 **U. Haagerup, H. P. Rosenthal, and F. A. Sukochev,** Banach embedding properties of non-commutative L^p-spaces, 2003
775 **P. Lochak, J.-P. Marco, and D. Sauzin,** On the splitting of invariant manifolds in multidimensional near-integrable Hamiltonian systems, 2003
774 **Kai A. Behrend,** Derived ℓ-adic categories for algebraic stacks, 2003
773 **Robert M. Guralnick, Peter Müller, and Jan Saxl,** The rational function analogue of a question of Schur and exceptionality of permutation representations, 2003
772 **Katrina Barron,** The moduli space of $N = 1$ superspheres with tubes and the sewing operation, 2003
771 **Shigenori Matsumoto,** Affine flows on 3-manifolds, 2003
770 **W. N. Everitt and L. Markus,** Elliptic partial differential operators and symplectic algebra, 2003
769 **Jie Wu,** Homotopy theory of the suspensions of the projective plane, 2003
768 **R. Höpfner and E. Löcherbach,** Limit theorems for null recurrent Markov processes, 2003
767 **Po Hu,** S-modules in the category of schemes, 2003

For a complete list of titles in this series, visit the
AMS Bookstore at **www.ams.org/bookstore/**.